STATICS STUDY PACK

CHAPTER REVIEWS, FREE-BODY DIAGRAM WORKBOOK, COMPANION WEBSITE

ENGINEERING MECHANICS

STATICS

THIRTEENTH EDITION

STATICS STUDY PACK

CHAPTER REVIEWS, FREE-BODY DIAGRAM WORKBOOK, COMPANION WEBSITE

ENGINEERING MECHANICS

STATICS

THIRTEENTH EDITION

R. C. HIBBELER

PEARSON

Upper Saddle River Boston Columbus San Francisco New York
Indianapolis London Toronto Sydney Singapore Tokyo Montreal Dubai
Madrid Hong Kong Mexico City Munich Paris Amsterdam Cape Town

Vice President and Editorial Director, ECS: Marcia Horton
Acquisitions Editor: Norrin Dias
Editorial Assistant: Sandra Rodriguez
Managing Editor: Scott Disanno
Art Director, Interior and Cover Designer: Kenny Beck
Art Editor: Gregory Dulles
Media Editor: Daniel Sandin
Operations Specialist: Lisa McDowell
Senior Marketing Manager: Tim Galligan
Marketing Assistant: Jon Bryant

About the Cover: Outdoor shot of electric train/© Nick M. Do / iStockphoto.com

---

Published by Pearson Prentice Hall
Pearson Education, Inc.
Upper Saddle River, New Jersey 07458

10 9 8 7 6 5 4 3 2 1

ISBN-10: 0-13-291556-1
ISBN-13: 978-0-13-291556-4

Contents

What's in This Package

The Statics Study Pack was designed to help students improve their study skills. It consists of three study components—a chapter-by-chapter review, a free-body diagram workbook, and an access code for the Companion Website.

- **Chapter-by-Chapter Review and Free-Body Diagram Workbook**—Prepared by Peter Schiavone of the University of Alberta. This resource contains chapter-by-chapter *Statics* review, including key points, equations, and check up questions. The Free-Body Diagram workbook steps students through numerous free-body diagram problems that include full explanations and solutions.
- **Companion Website**—Located at *www.prenhall.com/hibbeler*, the Companion Website includes the following resources:
    - **Video Solutions** — Complete, step-by-step solution walkthroughs of representative homework problems from each section in the Statics text. Developed by Professor Edward Berger of University of Virginia, Video Solutions offer:
        - **Fully-worked Solutions** — Showing every step of representative homework problems, to help students make vital connections between concepts.
        - **Self-paced Instruction** — Students can navigate each problem and select, play, rewind, fast-forward, stop, and jump-to sections within each problem's solution.
        - **24/7 Access** — Help whenever students need it with over 20 hours of helpful review.
    - **1000 Supplemental *Statics* and *Dynamics* Problems** — A self-study resource, these supplemental problems have complete solutions provided.
    - **MATLAB and Mathcad Tutorials** — Focused on using MATLAB and Mathcad in Engineering Mechanics, these tutorials are keyed to the text.

To log in to the Companion Website, follow the instructions on the access card included with this study pack.

Preface

This supplement is divided into two parts. Part I provides a section-by-section, chapter-by-chapter summary of the key concepts, principles and equations from R. C. Hibbeler's text, *Engineering Mechanics: Statics*, Thirteenth Edition. Part II is a workbook which explains how to draw and use free-body diagrams when solving problems in *Statics*.

Part I: Chapter-by-Chapter Summaries

This part of the supplement provides a section-by-section, chapter-by-chapter summary of the key concepts, principles and equations from R. C. Hibbeler's text, *Engineering Mechanics: Statics*, Thirteenth Edition. We follow the same section and chapter order as that used in the text and summarize important concepts from each section in easy-to-understand language. We end each chapter summary with a simple set of review questions designed to see if the student has understood the key concepts and chapter objectives.

This section of the supplement will be useful both as a quick reference guide for important concepts and equations when solving problems in, for example, homework assignments or laboratories and also as a handy review when preparing for any quiz, test, or examination.

Part II: Free-Body Diagram Workbook

A thorough understanding of how to draw and use a free-body diagram is absolutely essential when solving problems in mechanics.

This workbook consists mainly of a collection of problems intended to give the student practice in drawing and using free-body diagrams when solving problems in *Statics*.

All the problems are presented as *tutorial* problems with the solution only partially complete. The student is then expected to complete the solution by "filling in the blanks" in the spaces provided. This gives the student the opportunity to *build free-body diagrams in stages* and extract the relevant information from them when formulating equilibrium equations. Earlier problems provide students with partially drawn free-body diagrams and lots of hints to complete the solution. Later problems are more advanced and are designed to challenge the student more. The complete solution to each problem can be found on the back of the page. The problems are chosen from two-dimensional theories of particle and rigid body mechanics. Once the ideas and concepts developed in these problems have been understood and practiced, the student will find that they can be extended in a relatively straightforward manner to accommodate the corresponding three-dimensional theories.

The workbook begins with a brief primer on free-body diagrams: where they fit into the general procedure of solving problems in mechanics and why they are so important. Next follows a few examples to illustrate ideas and then the workbook problems.

For best results, the student should read the primer and then, beginning with the simpler problems, try to complete and understand the solution to each of the subsequent problems. The student should avoid the temptation to immediately look at the completed solution on the back of the page. This solution should be accessed only as a last resort (after the student has struggled to the point of giving up), or to check the student's own solution after the fact. The idea behind this is very simple: *we learn most when we* **do** *the thing we are trying to learn*—reading through someone else's solution is not the same as actually working through the problem. In the former, the student gains *information*, in the latter the student gains *knowledge*. For example, how many people learn to swim or drive a car by reading an instruction manual?

Consequently, since the workbook is based on **doing**, the student who persistently solves the problems in the workbook will ultimately gain a thorough, usable knowledge of how to draw and use free-body diagrams.

PETER SCHIAVONE

STATICS STUDY PACK

CHAPTER REVIEWS, FREE-BODY DIAGRAM WORKBOOK, COMPANION WEBSITE

ENGINEERING MECHANICS

STATICS

THIRTEENTH EDITION

PART I

Section-By-Section, Chapter-By-Chapter Summaries with Review Questions and Answers

1

General Principles

MAIN GOALS OF THIS CHAPTER:

- To introduce the basic ideas of *Mechanics.*
- To give a concise statement of Newton's laws of motion and gravitation.
- To review the principles for applying the SI system of units.
- To examine standard procedures for performing numerical calculations.
- To outline a general guide for solving problems.

1.1 MECHANICS

Mechanics is that branch of the physical sciences concerned with the behavior of bodies subjected to the action of forces. The subject of mechanics is divided into two parts:

- *statics*—the study of objects in equilibrium (objects either at rest or moving with a constant velocity).
- *dynamics*—the study of objects with accelerated motion.

Although statics can be considered as a special case of dynamics (in which the acceleration is zero), it deserves special treatment since many objects are designed with the intention that they remain in equilibrium.

1.2 FUNDAMENTAL CONCEPTS

BASIC QUANTITIES

- **Length, time, mass, force**

IDEALIZATIONS

Mathematical models or idealizations are used in mechanics to simplify the theory. The more common ones, in order of sophistication, are:

- **Particle**—a *particle* has a mass but a size that can be neglected i.e., the geometry of the body is ignored. A particle is often represented by a *point* in space.
- **Rigid Body**—a *rigid body* has a mass and a size (shape) but it is assumed that any changes in shape can be neglected i.e., the geometry of the body *is* taken into account but any deformations (changes in shape) are ignored. Consequently, the material properties of the body can be ignored. A rigid body is often represented as a collection of particles in which all the particles remain at a fixed distance from each other before and after applying a load.
- **Deformable or Elastic Body**—a deformable body has a mass, a size (shape) and the deformations (changes in shape) of the body are taken into account. Hence the material properties of the body must be considered in describing the behavior of the body.
- **Concentrated Force**—A concentrated force represents the effect of a loading which is assumed to act at a point on a body. This idealization requires that the area over which the load is applied is very small compared to the overall size of the body e.g., contact force between wheel and ground.

NEWTON'S THREE LAWS OF MOTION

Newton's laws apply to the motion of a particle as measured from a nonaccelerating (inertial) reference frame.

- **First Law**—a particle originally at rest or moving in a straight line with constant velocity, will remain in this state provided the particle is not subjected to an unbalanced force.
- **Second Law**—a particle acted upon by an unbalanced force **F** experiences an acceleration **a** that has the same direction as the force and a magnitude directly proportional to the force i.e.

$$\mathbf{F} = m\mathbf{a}$$

- **Third Law**—The mutual forces of action and reaction between two particles are equal, opposite and collinear.

NEWTON'S LAW OF GRAVITATIONAL ATTRACTION

$$F = G\frac{m_1 m_2}{r^2}$$

F = force of gravitation between two particles

G = universal constants of gravitation

m_1, m_2 = mass of each of the two particles

r = distance between the two particles

MASS AND WEIGHT

- **Mass** is a (scalar) property of matter that does not change from one location to another. In other words, mass is an *absolute* quantity
- **Weight** is a *force* (and hence has a magnitude and direction) which refers to the gravitational attraction of the earth on a quantity of mass m. Weight is not an absolute quantity. Its magnitude depends on the elevation at which the mass is located. We write the magnitude of weight as $W = mg$ where g is termed the acceleration due to gravity.

1.3 UNITS OF MEASUREMENT

The four basic quantities *force, mass, length* and *time* are related by Newton's 2nd law. Hence, the units used to define these quantities are not independent i.e., three of the four units are called *base units* (arbitrarily defined) and the fourth unit a *derived unit* (derived from Newton's 2nd law).

SI UNITS (INTERNATIONAL SYSTEM OF UNITS)

- In the SI system, the unit of force, the *newton*, is a derived unit. The meter, second and kilogram are base units.
- One *newton* is equal to a force required to give one kilogram of mass an acceleration of 1 m/s^2.
- In newtons, the weight of a body has magnitude

$$W = mg \text{ where } g = 9.81 \text{ m/s}^2.$$

US CUSTOMARY

- In the US Customary system, the unit of mass, the *slug*, is a derived unit. The foot, second and pound are base units.
- One *slug* is equal to the amount of matter accelerated at 1 ft/s^2 when acted upon by a force of 1 lb.
- In slugs, the mass of a body is given by

$$m = \frac{W}{g} \text{ where } g = 32.2 \text{ ft/s}^2.$$

The following table summarizes the two systems of units.

Name	Length	Time	Mass	Force
International System (SI)	meter (m)	second (s)	kilogram (kg)	newton* $\left(N = \frac{\text{kg}\cdot\text{m}}{\text{s}^2}\right)$
U.S. Customary (FPS)	foot (ft)	second (s)	slug* $= \left(\frac{\text{lb}\cdot\text{s}^2}{\text{ft}}\right)$	pound (lb)

**Derived Unit*

CONVERSION OF UNITS

The following table provides a set of direct conversion factors between FPS and SI units for the basic quantities. Note also that in the FPS system

- 1 ft = 12 in. (inches).
- 5280 ft = 1 mi. (mile).
- 1000 lb = 1 kip (1 kilo-pound).
- 2000 lb = 1 ton.

Quantity	Unit (FPS)	Equals	Unit (SI)
Force	lb		4.4482 N
Mass	slug		14.5938 kg
Length	ft		0.3048 m

1.4 THE INTERNATIONAL SYSTEM OF UNITS

PREFIXES

When a numerical quantity is either very large or very small, the units used to define its size may be modified by using a prefix. For example:

	Exponential Form	Prefix	SI Symbol
1 000 000 000	10^9	giga	G
1 000 000	10^6	mega	M
1 000	10^3	kilo	k
0.001	10^{-3}	milli	m
0.000 001	10^{-6}	micro	μ
0.000 000 001	10^{-9}	nano	n

RULES FOR USE

You should know the rules for the proper use of the various SI symbols. These are used extensively in engineering practice throughout the world.

1.5 NUMERICAL CALCULATIONS

It is important that the numerical answers to any problem encountered in engineering practice be reported with both justifiable accuracy and appropriate significant figures.

DIMENSIONAL HOMOGENEITY

Each term in any equation used to describe a physical process must be expressed in the same units i.e., the terms must be *dimensionally homogeneous.* Algebraic manipulations of an equation can be checked, in part, by verifying that the equation remains dimensionally homogeneous.

SIGNIFICANT FIGURES

The accuracy of a number is specified by the number of significant figures it contains. A *significant figure* is any digit, including a zero, provided it is not used to specify the location of the decimal point for the number. For example, 0.00546 and 2500 expressed to three significant figures would be $5.46 \times (10^{-3})$ and $2.50 \times (10^3)$, respectively *(engineering notation)*.

ROUNDING OFF NUMBERS

For numerical calculations, the accuracy of the solution of a problem (generally) can never be better than the accuracy of the problem data. Consequently, a calculated result should always be *rounded off* to an appropriate number of significant figures. To convey appropriate accuracy, there are rules for rounding off numbers. You should know these.

CALCULATIONS

Perform numerical calculations to *several significant figures* and then report the final answer to *three* significant figures.

1.6 GENERAL PROCEDURE FOR ANALYSIS

- The most effective way to learn engineering mechanics is to *solve problems.*
- You must present your work in a logical and orderly manner as follows:

- Read the problem carefully and try to establish a link between the actual physical situation and the appropriate part of the theory studies.
- Draw any necessary diagrams and tabulate the problem data.
- Apply the relevant principles.
- Solve the necessary equations algebraically, as far as practical, then, making sure they are dimensionally homogeneous, use a consistent set of units and complete the solution numerically. Report the answer with no more significant figures than the accuracy of the given data.
- Study the answer and see if it *makes sense* physically—in the context of the physical problem.

HELPFUL TIPS AND SUGGESTIONS

- The *language* of engineering mechanics is *mathematics*. Consequently, make sure you review/re-read the necessary mathematical notation/concepts *as they arise* in your mechanics course (trying to review all of the necessary mathematics *at once* is not recommended—there's just too much to digest at one time). You should aim to achieve *fluency* in basic mathematical techniques/notation so that your learning of mechanics is not distracted by trying to remember things which your instructor *assumes* you know e.g., how to solve linear systems of algebraic equations, how to perform basic vector algebra, differentiation and integration etc.
- *Remember* that in solving problems from engineering mechanics you are solving real practical problems and producing real data with physical significance. Thus, you are responsible for making sure your results are correct, consistent and well-presented. Get into the habit of doing this *now* so that it will become second nature by the time you graduate. In the world of professional engineering you have a responsibility to your profession and to the many people that will use the product you will help to design, manufacture or implement.

REVIEW QUESTIONS: TRUE OR FALSE[1]?

1. The subject called *Statics* studies only bodies which are at rest.
2. A *particle* has a mass but negligible shape/size.
3. A rigid body has a mass but negligible shape/size.
4. Newton's three laws of motion can be proved mathematically.
5. Weight is a property of matter that does not change from one location to another.
6. In the *SI* system of units, the newton is a derived unit.
7. When performing numerical calculations, the final answer should be reported to *three* significant figures.
8. In an equation it's permitted to have different terms expressed in different units. This is referred to as dimensionally inhomogeneous.

[1] 1.F 2.T 3. F 4. F 5. F 6. T 7. T 8. F

2

Force Vectors

MAIN GOALS OF THIS CHAPTER

In this chapter we define scalars, vectors and vector operations and use them to analyze forces acting on objects. Specifically:

- To show how to add forces and resolve them into components.
- To express force and position in Cartesian vector form.
- To explain how to determine a vector's magnitude and direction.
- To introduce the dot product and use it to find the angle between two vectors or the projection of one vector onto another.

2.1 SCALARS AND VECTORS

Most of the physical quantities in mechanics can be represented by either *scalars* or *vectors*:

- A *scalar* is a real number e.g., mass, time, volume and length are represented by scalars.
- A *vector* has both magnitude and direction e.g., force, velocity and acceleration are vectors.

2.2 VECTOR OPERATIONS

MULTIPLICATION OR DIVISION OF A VECTOR BY A SCALAR

- The product of a vector $\mathbf{A}$ and a scalar a is a vector $a\mathbf{A}$ with magnitude $|a\mathbf{A}| = |a|\,|\mathbf{A}|$. The direction is the same as that of $\mathbf{A}$ if a is positive and opposite to that of $\mathbf{A}$ if a is negative.

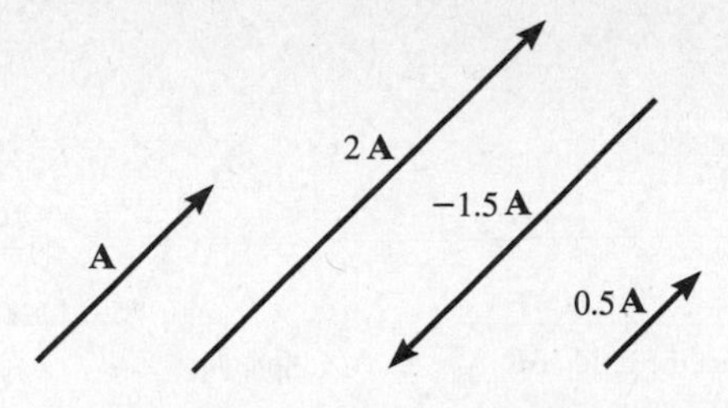

Scalar Multiplication and Division

VECTOR ADDITION

- Two vectors **A** and **B** can be added to form a *resultant* vector $\mathbf{R} = \mathbf{A} + \mathbf{B}$ by using the parallelogram law. If the two vectors are *collinear* (both vectors have the same line of action), the resultant is formed by an algebraic or scalar addition.

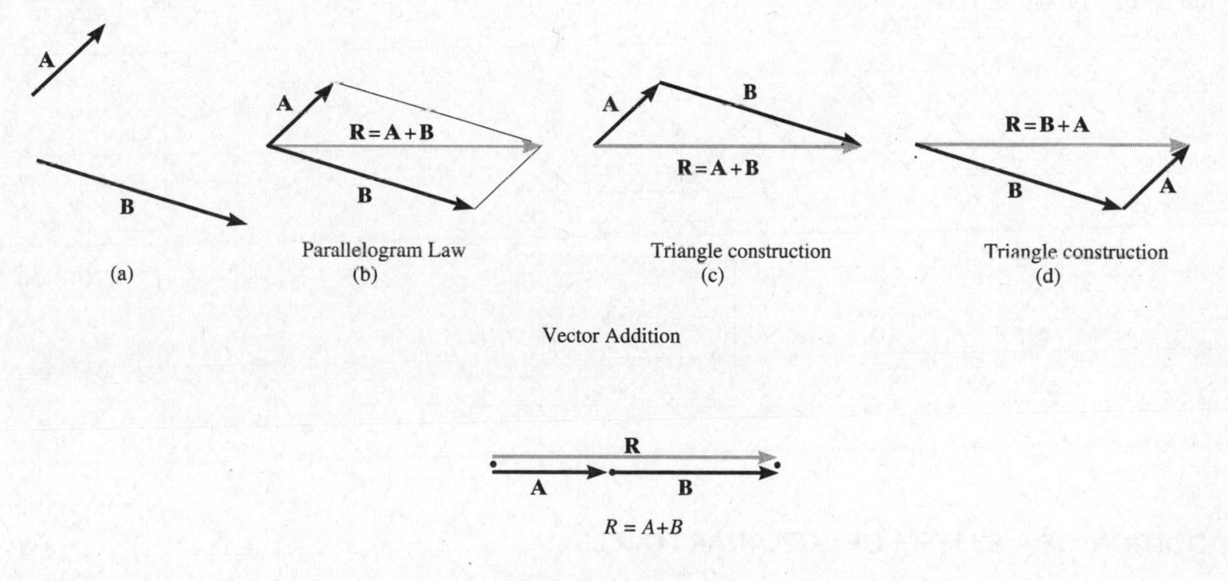

Addition of collinear vectors

RESOLUTION OF A VECTOR

- A vector may be resolved into *components* having known lines of action by using the parallelogram law.

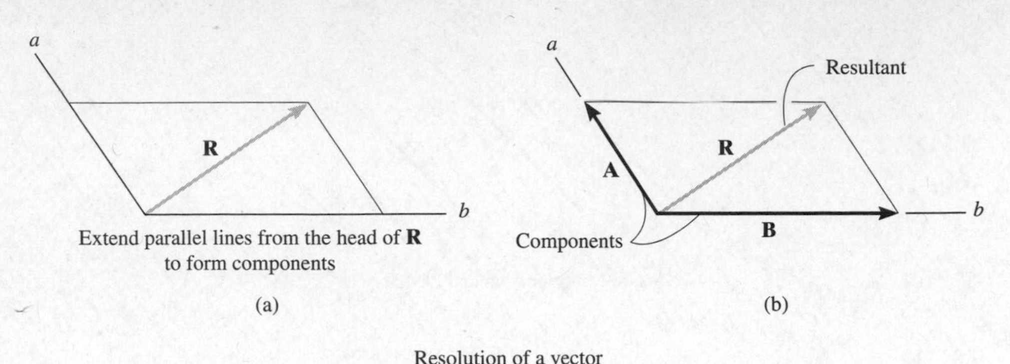

Resolution of a vector

2.3 VECTOR ADDITION OF FORCES

- A force is a *vector* quantity since it has a specified magnitude and direction. Consequently, forces are added together or resolved into components using the rules of vector algebra.
- Two common problems in statics involve either finding the resultant force given its components or resolving a known force into components.
- Often the magnitude of a resultant force can be determined from the law of cosines, while its direction is determined from the law of sines:

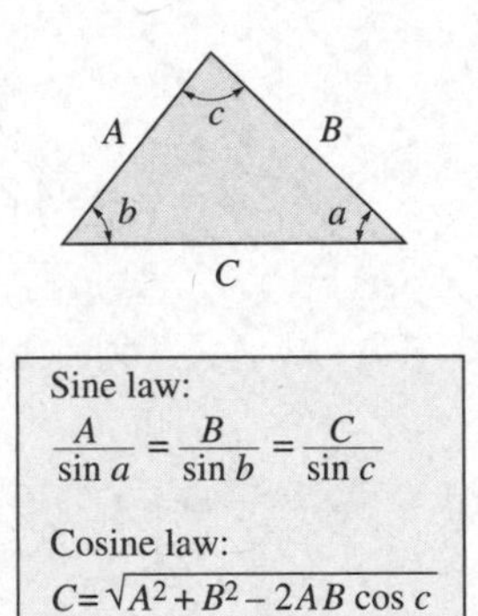

Sine law:

$$\frac{A}{\sin a} = \frac{B}{\sin b} = \frac{C}{\sin c}$$

Cosine law:

$$C = \sqrt{A^2 + B^2 - 2AB\cos c}$$

2.4 ADDITION OF A SYSTEM OF COPLANAR FORCES

In the plane, a force can be resolved into two rectangular components. There are two separate notations for doing this:

- *Scalar Notation*—we write the force **F** as (F_x, F_y) where F_x and F_y are the scalar components of the force **F** in the directions of the positive x- and y-axes, respectively. If F_x and F_y are negative, it means that $|F_x|$ and $|F_y|$ are directed along the *negative* x- and y-axes, respectively.
- *Cartesian Vector Notation*—we write the force **F** as

$$\mathbf{F} = F_x\mathbf{i} + F_y\mathbf{j},$$

where **i** and **j** represent the positive directions of the x- and y-axes, respectively.

COPLANAR FORCE RESULTANTS

- The resultant of several coplanar forces can easily be determined if an x, y-coordinate system is established and the forces are resolved along the axes. For example,

$$\begin{aligned} \mathbf{F}_1 &= F_{1x}\mathbf{i} + F_{1y}\mathbf{j}, \\ \mathbf{F}_2 &= F_{2x}\mathbf{i} + F_{2y}\mathbf{j}, \\ \mathbf{F}_3 &= F_{3x}\mathbf{i} + F_{3y}\mathbf{j}, \end{aligned}$$

  then the resultant is given by

$$\begin{aligned} \mathbf{F}_R &= \mathbf{F}_1 + \mathbf{F}_2 + \mathbf{F}_3 \\ &= \left(F_{1x} + F_{2x} + F_{3x}\right)\mathbf{i} + \left(F_{1y} + F_{2y} + F_{3y}\right)\mathbf{j} \\ &= \left(F_{Rx}\right)\mathbf{i} + \left(F_{Ry}\right)\mathbf{j}. \end{aligned}$$

- In the general case, the x and y components of the resultant of any number of coplanar forces can be represented symbolically by the algebraic sum of the x and y components of all the forces i.e.

$$\begin{aligned} F_{Rx} &= \sum F_x, \\ F_{Ry} &= \sum F_y. \end{aligned}$$

- The magnitude and direction of the resultant force are given by:

$$\begin{aligned} |\mathbf{F}_R| = F_R &= \sqrt{F_{Rx}^2 + F_{Ry}^2}, \\ \theta &= \tan^{-1}\left|\frac{F_{Ry}}{F_{Rx}}\right|, \end{aligned} \tag{2.0}$$

  respectively.

2.5 CARTESIAN VECTORS

- A Cartesian coordinate system is often used to solve problems in three dimensions. The coordinate system is *right-handed* which means that the thumb of the right hand points in the direction of the positive z-axis when the right hand fingers are curled about this axis and directed from the positive x toward the positive y-axis.

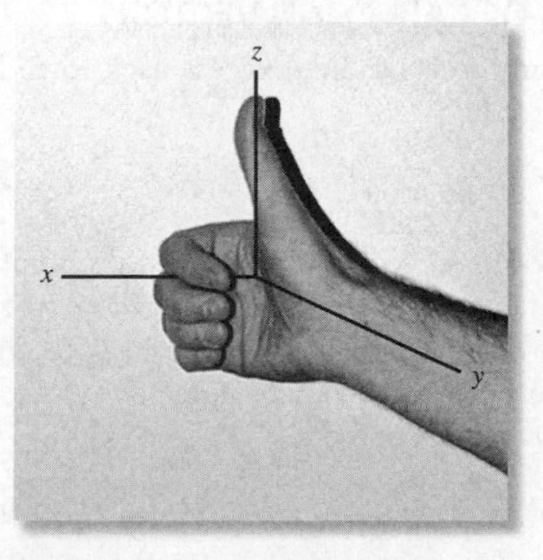

Right-handed coordinate system.

- The unit vector from a vector **A** is given by $\frac{\mathbf{A}}{A}$ where $A \neq 0$ is the magnitude of vector **A**. The unit vector is dimensionless and defines the direction of vector **A**.
- The positive directions of the x, y, z axes are defined by the Cartesian unit vectors **i**, **j**, **k**, respectively. Consequently, any vector **A** with scalar components A_x, A_y and A_z can be written in the Cartesian vector form

$$\mathbf{A} = A_x\mathbf{i} + A_y\mathbf{j} + A_z\mathbf{k}. \tag{2.1}$$

- The magnitude of vector **A** is given by

$$|\mathbf{A}| = A = \sqrt{A_x^2 + A_y^2 + A_z^2}. \tag{2.2}$$

- The direction of vector **A** is defined by the angles α, β and γ measured between the tail of **A** and the positive x, y, z axes located at the tail of **A**.

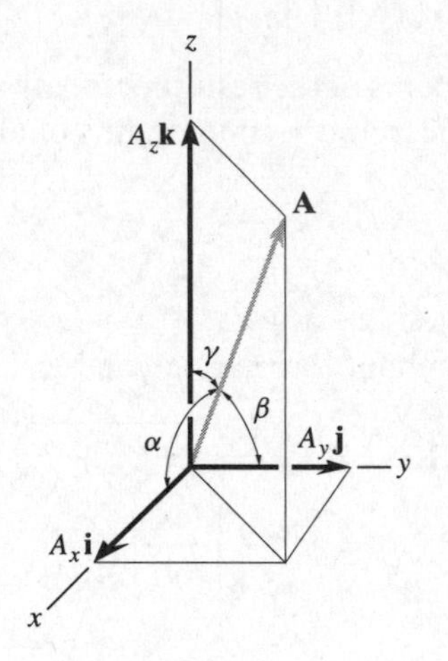

- The angles α, β and γ are found from their *direction cosines*

$$\cos\alpha = \frac{A_x}{A}, \cos\beta = \frac{A_y}{A}, \cos\gamma = \frac{A_z}{A}, \quad \cos^2\alpha + \cos^2\beta + \cos^2\gamma = 1. \tag{2.3}$$

This means that only two of the angles α, β and γ have to be specified—the third can be found from $\cos^2\alpha + \cos^2\beta + \cos^2\gamma = 1$.

2.6 ADDITION AND SUBTRACTION OF CARTESIAN VECTORS

- To find the resultant of a concurrent force system, express each force as a Cartesian vector and add the **i**, **j**, **k** components of all the forces in the system.

2.7 POSITION VECTORS

- The position vector **r** is defined as a fixed vector which locates a point in space relative to another point. For example, from the origin of coordinates O, the point in space $P\,(x, y, z)$ has position vector $\mathbf{r} = x\mathbf{i}+y\mathbf{j}+z\mathbf{k}$.
- More generally, the position vector may be directed from point A to point B in space. In this case, the position vector is again denoted by **r** (or sometimes $\mathbf{r}_{AB}$) and is given by

$$\mathbf{r}_{AB} = \mathbf{r}_B - \mathbf{r}_A \tag{2.4}$$

where $\mathbf{r}_B$ and $\mathbf{r}_A$ are the position vectors of A and B from the origin of coordinates O. For example, if $A\,(x_A, y_A, z_A)$ and $B\,(x_B, y_B, z_B)$ then

$$\mathbf{r}_A = x_A\mathbf{i}+y_A\mathbf{j}+z_A\mathbf{k}, \quad \mathbf{r}_B = x_B\mathbf{i}+y_B\mathbf{j}+z_B\mathbf{k}, \quad \mathbf{r}_{AB} = (x_B - x_A)\mathbf{i} + (y_B - y_A)\mathbf{j} + (z_B - z_A)\mathbf{k}.$$

2.8 FORCE VECTOR DIRECTED ALONG A LINE

- A force **F** (with magnitude F) acting in the direction of a line represented by a position vector **r** can be written in the form

$$\mathbf{F} = F\left(\frac{\mathbf{r}}{r}\right) = F\mathbf{u}$$

where $\mathbf{u} = \frac{\mathbf{r}}{r}$ is a unit vector representing the direction of the line.

2.9 DOT PRODUCT

- The dot product is used to determine
  - The angle between two vectors.
  - The projection of a vector in a specified direction.
- The dot product of two vectors **A** and **B** is defined as

$$\mathbf{A} \cdot \mathbf{B} = A_x B_x + A_y B_y + A_z B_z = AB\cos\theta, \tag{2.5}$$

where A and B are the magnitudes of **A** and **B**, respectively, and θ is the angle between the tails of **A** and **B**. Consequently,

$$\theta = \cos^{-1}\left(\frac{\mathbf{A} \cdot \mathbf{B}}{AB}\right).$$

- The dot product is commutative ($\mathbf{A} \cdot \mathbf{B} = \mathbf{B} \cdot \mathbf{A}$), and distributive $\mathbf{A} \cdot (\mathbf{B} + \mathbf{D}) = \mathbf{A} \cdot \mathbf{B} + \mathbf{A} \cdot \mathbf{D}$. Also, if $a \in \mathbb{R}$:

$$a(\mathbf{A} \cdot \mathbf{B}) = (a\mathbf{A}) \cdot \mathbf{B} = \mathbf{A} \cdot (a\mathbf{B}) = (\mathbf{A} \cdot \mathbf{B})a. \tag{2.6}$$

- In some engineering applications, you must resolve a vector into components which are parallel and perpendicular (normal) to a given line (direction). The component of vector **A** in the direction specified by the unit vector **u** is given by

$$A_{\parallel} = \mathbf{A} \cdot \mathbf{u} = A\cos\theta. \tag{2.7}$$

This component is also referred to as the scalar projection of **A** onto the line with direction **u** or the component of vector **A** parallel to a line with direction **u**. Clearly, the vector $\mathbf{A}_{\parallel}$ is defined by $\mathbf{A}_{\parallel} = (A_{\parallel})\,\mathbf{u}$.

- Once the parallel component has been determined, we can determine the component of **A** perpendicular (or normal) to a line with direction **u** by

$$A_{\perp} = \sqrt{A^2 - A_{\parallel}^2}. \tag{2.8}$$

- Clearly, in terms of vectors,

$$\mathbf{A} = \mathbf{A}\| + \mathbf{A}\perp.$$

HELPFUL TIPS AND SUGGESTIONS

- Be aware of the differences between vectors and scalars. For example, force, velocity and acceleration are *vectors* while speed, time and distance are *scalars*. If you are asked to find a vector (e.g., a force) you must report *both* magnitude and direction.
- Vector operations are essential in describing the basic principles of mechanics. Make sure you take the time to review basic vector algebra. It doesn't take long but the payoff (in terms of your effectiveness in mechanics) is significant.

REVIEW QUESTIONS

1. How are the scalar components of a vector defined in terms of a Cartesian coordinate system?
2. If you know the scalar components of a vector, how can you determine its magnitude and direction?
3. Suppose you know the coordinates of two points A and B. How do you determine the scalar components of the position vector of point B relative to point A?
4. How do you identify a right-handed coordinate system?
5. What are the direction cosines of a vector? If you know them, how do you determine the components of the vector?
6. What is the definition of the dot product? Is the dot product a vector or a scalar?
7. If the dot product of two vectors is zero, what does that mean?
8. If you know the components of two vectors **A** and **B**, how can you determine their dot product?
9. Simplify
   i. $\mathbf{A} \cdot (\mathbf{B} + \mathbf{D})$
   ii. $(a\mathbf{A}) \cdot \mathbf{B}$ where a is a scalar.
10. How can you use the dot product to determine the components of a vector parallel and perpendicular to a line?

3

Equilibrium of a Particle

MAIN GOALS OF THIS CHAPTER

In this chapter we:

- Introduce the concept of the free-body diagram for an object modelled as a particle (an object with mass but negligible shape/size—henceforth referred to simply as a *particle*).
- Show how to solve particle equilibrium problems using the equations of equilibrium.

3.1 CONDITION FOR THE EQUILIBRIUM OF A PARTICLE

A particle is in *equilibrium* provided it is at rest if originally at rest (*static equilibrium*) or has a constant velocity if originally in motion.

- To maintain equilibrium it is necessary and sufficient that the *resultant force* acting on a particle be equal to zero. In terms of Newton's laws of motion, this is expressed mathematically as:

$$\sum \mathbf{F} = \mathbf{0} \tag{3.0}$$

where $\sum \mathbf{F}$ is the vector sum of all forces acting on the particle.

3.2 THE FREE-BODY DIAGRAM

To apply the equation of equilibrium (3.0), we must account for all the known and unknown forces ($\sum \mathbf{F}$) which act on the particle. The easiest way to do this is to draw a *free-body diagram*.

- A free-body diagram is simply a sketch which shows the particle 'free' from its surroundings with *all* the forces that act *on* it. There are three main steps:
    - ◆ **Draw Outlined Shape**. Imagine the particle to be isolated or cut 'free' from its surroundings by drawing its outlined shape.
    - ◆ **Show all Forces**. Indicate on this sketch *all* the forces that act *on the particle*—it may help to carefully trace around the particle's boundary, noting each force acting.
    - ◆ **Identify Each Force**. The forces which are known should be labeled with their proper magnitudes and directions. Letters are used to represent the magnitudes and directions of forces that are unknown.

* **Connections**—there are two types of connections often encountered in particle equilibrium problems:
  (a) *Springs*—The magnitude of force exerted on a linear elastic spring with stiffness k, deformed a distance s measured from its unloaded position is

$$F = ks. \tag{3.1}$$

  Here s is determined from the difference in the spring's deformed length and its undeformed length.
  (b) *Cables and Pulleys*—Assume cables (or cords) have negligible weight and cannot stretch. Also, a cable can support only tension which always acts in the direction of the cable.

There are several examples and practice problems, as well as much more on drawing free-body diagrams in Part II of this study pack.

3.3 COPLANAR FORCE SYSTEMS

Coplanar force equilibrium problems for a particle can be solved using the following procedure.

1. Free-Body Diagram

- Establish the x, y axes in any suitable orientation.
- Label all the known and unknown force magnitudes and directions on the diagram.
- The sense of a force having an unknown magnitude can be assumed.

2. Equations of Equilibrium

- Resolve each force into its **i** (x) and **j** (y) components and apply the *scalar equations of equilibrium*

$$\sum F_x = 0, \quad \sum F_y = 0. \tag{3.2}$$

  (the algebraic sum of the x and y components of all the forces acting on the particle equal to zero).
- Components are positive if they are directed along a positive axis and negative if they are directed along a negative axis.
- If more than two unknowns exist and the problem involves a spring, apply $F = ks$ to relate the spring force to the deformation s of the spring.
- If the solution yields a negative result, this indicates the sense of the force is the reverse of that shown on the free-body diagram.

3.4 THREE-DIMENSIONAL FORCE SYSTEMS

Three-dimensional force equilibrium problems for a particle can be solved using the following procedure.

1. Free-Body Diagram

- Establish the x, y, z axes in any suitable orientation.
- Label all the known and unknown force magnitudes and directions on the diagram.
- The sense of a force having an unknown magnitude can be assumed.

2. Equations of Equilibrium

- When it's easy to do so, resolve each force into its $\mathbf{i}(x)$, $\mathbf{j}(y)$, and $\mathbf{k}(z)$ components and apply the *scalar equations of equilibrium*

$$\sum F_x = 0, \quad \sum F_y = 0, \quad \sum F_z = 0. \qquad (3.3)$$

(the algebraic sum of the x , y and z components of all the forces acting on the particle equal to zero).

- If the three-dimensional geometry appears difficult, then first express each force as a Cartesian vector and substitute these vectors into the *vector equation of equilibrium* (3.0)

$$\sum \mathbf{F} = \mathbf{0}$$

and then set the **i** , **j** and **k** components equal to zero.

- If the solution yields a negative result, this indicates the sense of the force is the reverse of that shown on the free-body diagram.

HELPFUL TIPS AND SUGGESTIONS

- Since we must account for *all the forces acting on the (object modelled as a) particle* when applying the equations of equilibrium, the importance of *first* drawing a free-body diagram cannot be over-emphasized.
- **One of the most common mistakes made in writing equilibrium conditions is forgetting to include all of the forces acting.** When drawn carefully, a free-body diagram will make it easier for you to identify *all* the forces acting.
- Use Part II of this supplement to get lots of practice in drawing free-body diagrams and applying the equations of equilibrium for a particle.

REVIEW QUESTIONS

1. What is meant by 'equilibrium of a particle'?
2. What do you know about the sum of the external forces acting on an object modelled as a particle in equilibrium?
3. What are the steps in drawing a free-body diagram?
4. What is a coplanar force system?
5. What is a three-dimensional system of forces?
6. What is the difference between equilibrium of coplanar and three-dimensional force systems?
7. What is the relation between the magnitude of the force exerted on a linear spring and the change in its length?
8. The following is the correct free-body diagram for the ring at E. True or False?

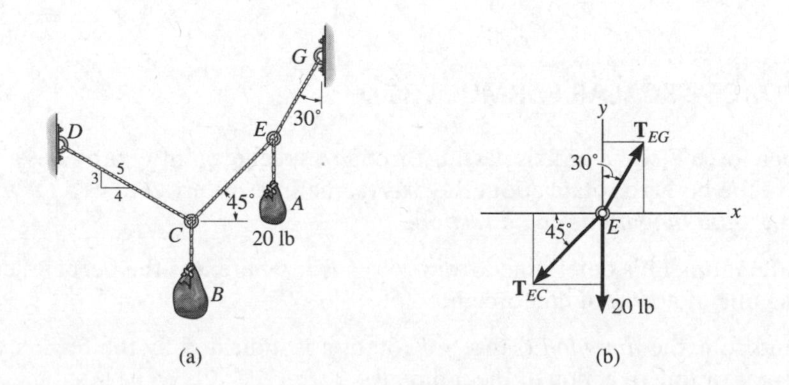

4

Force System Resultants

MAIN GOALS OF THIS CHAPTER:

- To discuss the concept of the *moment of a force* and show how to calculate it in two and three dimensions.
- To provide a method for finding the moment of a force about a specified axis.
- To define the moment of a couple.
- To present methods for determining the resultants of nonconcurrent force systems.
- To indicate how to reduce a simple distributed loading to a resultant force having a specified location.

4.1 MOMENT OF A FORCE—SCALAR FORMULATION

- The *moment* $\mathbf{M}_O$ of a force $\mathbf{F}$ about an axis passing through a specific point O provides a measure of the tendency of the force to cause the body to rotate about the axis (sometimes referred to as a *torque*). Clearly the moment is a *vector* and so has *both magnitude and direction*.
- The *magnitude* of the moment is determined from $M_0 = Fd$, where d is the perpendicular or shortest distance from point O to the line of action of the force $\mathbf{F}$.
- Using the right-hand rule, the *direction* (sense) of rotation is indicated by the fingers with the thumb directed along the moment axis or line of action of the moment.

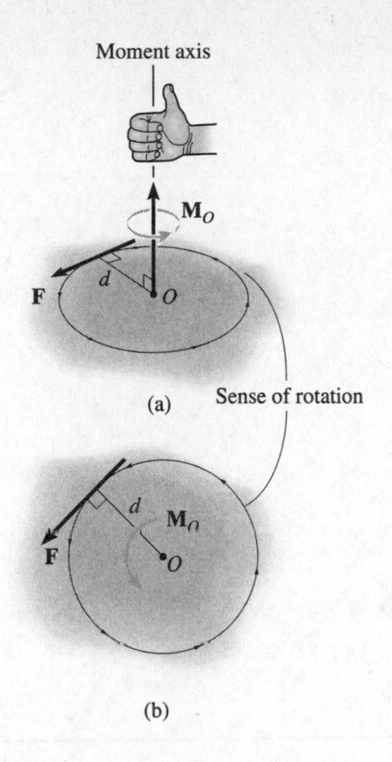

- If a system of forces lies in the $x-y$ plane, then the moment produced by each force about point O will be directed along the z-axis. The resultant moment $\mathbf{M}_{R_O}$ of the system can be determined by simply adding the moments of all the forces algebraically since all the moment vectors are *collinear* i.e.

$$\curvearrowleft +M_{R_O} = \sum F_d.$$

Here the counterclockwise curl written alongside the equation indicates that the moment of any force will be positive if it is directed along the z-axis, whereas a negative moment is directed along the z-axis.

4.2 CROSS PRODUCT

- The cross product of two vectors **A** and **B** yields a *vector* **C** written **C** = **A** × **B**.

  - The *magnitude* of vector **C** is given by $AB \sin\theta$ where θ is the angle between the tails of **A** and **B**.

  - Vector **C** has a direction which is perpendicular to the plane containing **A** and **B** such that **C** is specified by the right-hand rule i.e., curling the fingers of the right hand from vector **A** (cross) to vector **B**, the thumb then points in the direction of **C**.

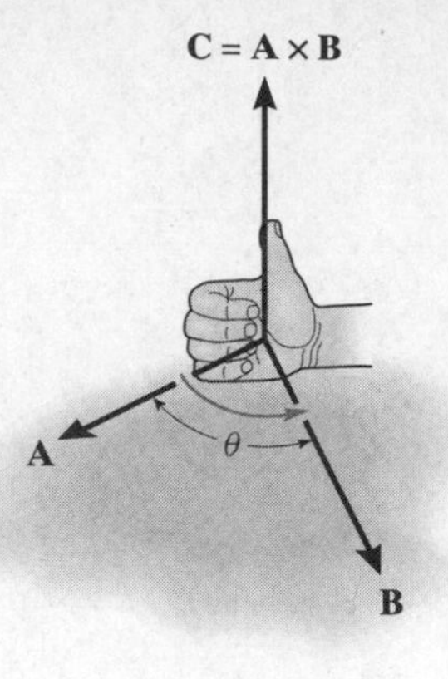

LAWS OF OPERATION

- $\mathbf{A} \times \mathbf{B} \neq \mathbf{B} \times \mathbf{A}$, rather $\mathbf{A} \times \mathbf{B} = -\mathbf{B} \times \mathbf{A}$.
- $a(\mathbf{A} \times \mathbf{B}) = (a\mathbf{A}) \times \mathbf{B} = \mathbf{A} \times (a\mathbf{B}) = (\mathbf{A} \times \mathbf{B})a$.
- $\mathbf{A} \times (\mathbf{B} + \mathbf{D}) = (\mathbf{A} \times \mathbf{B}) + (\mathbf{A} \times \mathbf{D})$.

CARTESIAN VECTOR FORMULATION

- To find the cross product of any two Cartesian vectors **A** and **B** we use the determinant

$$\mathbf{A} \times \mathbf{B} = \begin{vmatrix} \mathbf{i} & \mathbf{j} & \mathbf{k} \\ A_x & A_y & A_z \\ B_x & B_y & B_z \end{vmatrix}. \tag{4.0}$$

- The following useful results can be obtained by applying the right-hand rule and don't need to be memorized.

$\mathbf{i} \times \mathbf{j} = \mathbf{k}$,	$\mathbf{i} \times \mathbf{k} = -\mathbf{j}$,	$\mathbf{i} \times \mathbf{i} = \mathbf{0}$,
$\mathbf{j} \times \mathbf{k} = \mathbf{i}$,	$\mathbf{j} \times \mathbf{i} = -\mathbf{k}$,	$\mathbf{j} \times \mathbf{j} = \mathbf{0}$,
$\mathbf{k} \times \mathbf{i} = \mathbf{j}$,	$\mathbf{k} \times \mathbf{j} = -\mathbf{i}$,	$\mathbf{k} \times \mathbf{k} = \mathbf{0}$.

4.3 MOMENT OF A FORCE—VECTOR FORMULATION

In three-dimensions it is preferable to use the vector cross product to determine the moment:

- The moment $\mathbf{M}_O$ of a force **F** about the moment axis passing through point O and perpendicular to the plane containing O and **F** can be represented by

$$\mathbf{M}_O = \mathbf{r} \times \mathbf{F} = \begin{vmatrix} \mathbf{i} & \mathbf{j} & \mathbf{k} \\ r_x & r_y & r_z \\ F_x & F_y & F_z \end{vmatrix}, \tag{4.1}$$

where **r** represents a position vector drawn *from O to any point* lying on the line of action of **F**.

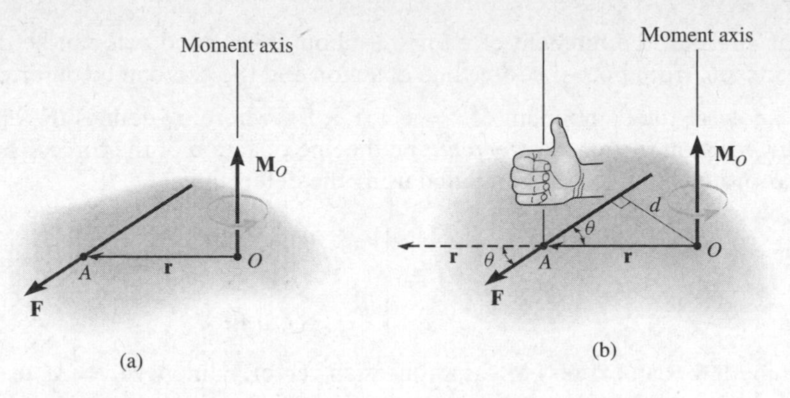

PRINCIPLE OF TRANSMISSIBILITY

- Since in (4.1), **r** can extend from O to *any* point on the line of action of **F**, **F** is a *sliding vector* and can act at *any* point along its line of action and create the *same moment* about point O.

RESULTANT MOMENT OF A SYSTEM OF FORCES

- If a body is acted upon by a system of n forces, the resultant moment about O is just the vector sum of the individual moments:

$$\mathbf{M}_{R_O} = \sum_{i=1}^{n} (\mathbf{r}_i \times \mathbf{F}_i).$$

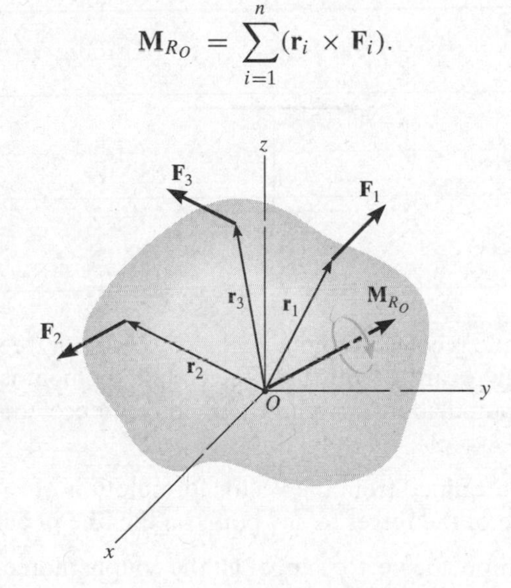

4.4 PRINCIPLE OF MOMENTS

- The *principle of moments* (Varignon's theorem) states that *the moment of a force about a point is equal to the sum of the moments of the force's components about the point.* This is particularly convenient since it is often easier to determine the moments of a force's components rather than the moment of the force itself (e.g., in two dimensions).

4.5 MOMENT OF A FORCE ABOUT A SPECIFIED AXIS

Recall that when the moment of a force is computed about a point, the moment and its axis are *always* perpendicular to the plane containing the force and the moment arm. In some problems it is important to find the *component* of this moment about a *specified axis* that passes through the point.

- In terms of a scalar analysis, the moment of a force **F** about a specified axis can be determined provided the perpendicular distance d_a from both the force line of action and the axis can be determined. Then $M_a = Fd_a$.
- In terms of vector analysis, the component $M_a = \mathbf{u}_a \cdot (\mathbf{r} \times \mathbf{F})$, where $\mathbf{u}_a$ defines the direction of the axis and **r** is directed from *any point* on the axis to *any point* on the line of action of the force. The quantity $\mathbf{u}_a \cdot (\mathbf{r} \times \mathbf{F})$ is called a triple scalar product and can be computed using the determinant

$$\mathbf{u}_a \cdot (\mathbf{r} \times \mathbf{F}) = \begin{vmatrix} u_{a_x} & u_{a_y} & u_{a_z} \\ r_x & r_y & r_z \\ F_x & F_y & F_z \end{vmatrix}$$

Once M_a is determined, we can express $\mathbf{M}_a$ as a Cartesian vector, namely $\mathbf{M}_a = M_a\mathbf{u}_a$.

- If M_a is calculated as a negative scalar then the sense of direction of $\mathbf{M}_a$ is opposite to $\mathbf{u}_a$.

4.6 MOMENT OF A COUPLE

- A couple is defined as two parallel forces that have the same magnitude, opposite directions, and are separated by a perpendicular distance d. Since the resultant force is zero, the only effect of a couple is to produce a rotation in a specified direction.

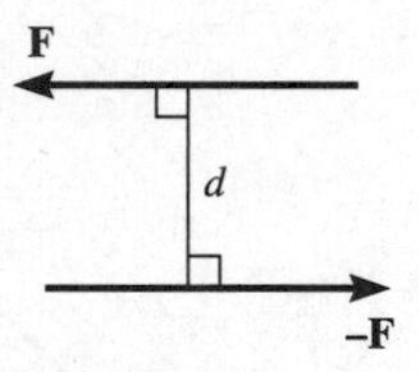

- The moment produced by a couple is called a *couple moment* which is a free vector and, as a result, it causes the same effect of rotation on a body regardless of where the couple moment is applied to the body. Consequently, the couple moment can be computed about *any* point. For convenience, this point is often chosen on the line of action of one of the forces in the couple.
- The couple moment is easily determined from the vector formulation $\mathbf{M} = \mathbf{r} \times \mathbf{F}$ where **r** is directed from *any point* on the line of action of one of the forces to any point on the line of action of the other force **F**.
- A *resultant couple moment* is simply the vector sum of all the couple moments of the system.

4.7 SIMPLIFICATION OF A FORCE AND COUPLE SYSTEM

A force has the effect of both translating and rotating a body and the amount by which it does so depends on where and how the force is applied. It is possible, however, to *replace* a system of forces and couple moments acting on a body with an *equivalent single* resultant force and couple moment acting at a specified point O. Here *equivalent* means that the system and the resultant each produce the same *external effects* of translation and rotation. There are two cases to consider:

- **Point O is on the Line of Action of the Force**—simply slide the force along its line of action to the point O.

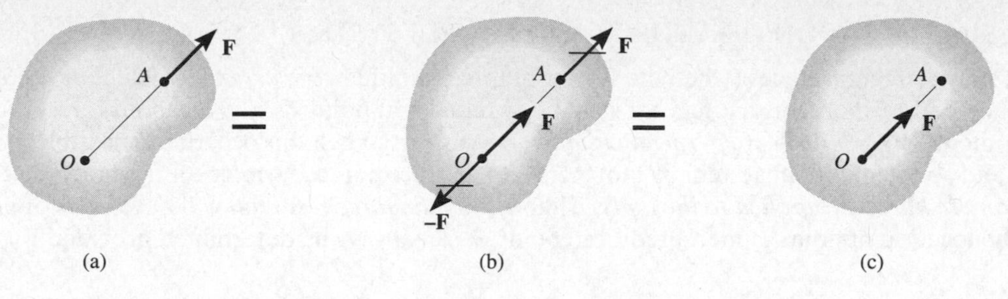

- **Point O is not on the Line of Action of the Force**—move the force to the point O and add a couple moment anywhere to the body.

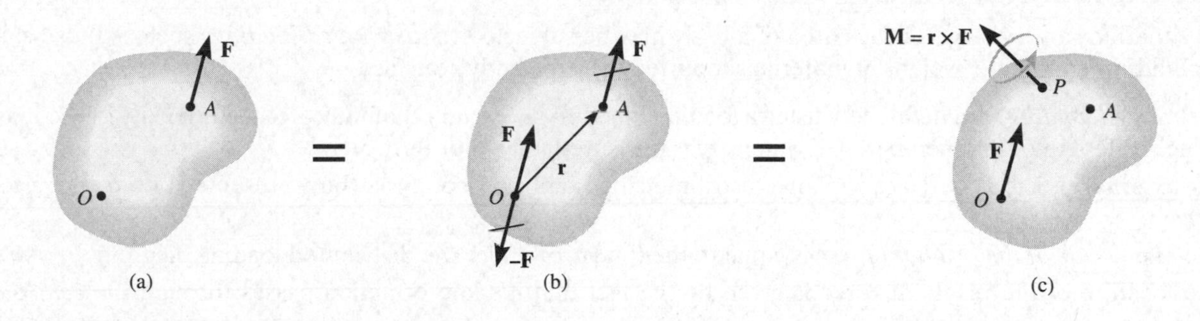

Next, we show how to determine the *equivalent resultants* mentioned above.

- To simplify any force and couple moment system to a resultant force acting at point O and a resultant couple moment we use the following equations

$$\mathbf{F}_R = \sum \mathbf{F}, \qquad (4.2)$$
$$(\mathbf{M}_R)_O = \sum \mathbf{M}_C + \sum \mathbf{M}_O$$

  The first equation states that the resultant force of the system is equivalent to the sum of all the forces. The second equation states that the resultant couple moment of the system is equivalent to the sum of all the couple moments $\sum \mathbf{M}_C$ plus the moments $\sum \mathbf{M}_O$ about point O of all the forces.

- The following tips may prove useful when applying equations (4.2):

  - Establish the coordinate axes with the origin located at point O and the axes having a selected orientation.
  - If the force system lies in the x–y plane, and any couple moments are perpendicular to this plane i.e. along the z-axis, the equations (4.2) reduce to the scalar equations:

$$(F_R)_x = \sum F_x,$$
$$(F_R)_y = \sum F_y$$
$$(M_R)_O = \sum M_C + \sum M_O.$$

4.8 FURTHER SIMPLIFICATION OF A FORCE AND COUPLE SYSTEM

- In certain special circumstances (when the system of forces is either *concurrent, coplanar or parallel*), the system of forces and couple moments acting on a rigid body reduces at point O to a resultant force $\mathbf{F}_R$ and a resultant couple moment $(\mathbf{M}_R)_O$ which are *perpendicular to one another*. When this occurs, it is possible to further simplify the force and couple moment system by moving $\mathbf{F}_R$ to another point P (on or off the body) so that *no resultant couple moment has to be applied to the body*. That is, *only the force resultant* will have to be applied to the body (at P). The location of point P measured from point O can always be determined provided $\mathbf{F}_R$ and $(\mathbf{M}_R)_O$ are known.
- **Reduction to a Wrench**—A general force and couple moment system acting on a body will reduce to a single resultant force $\mathbf{F}_R$ and a resultant couple moment $(\mathbf{M}_R)_O$ at O which are *not* perpendicular to one another. In this case the force and couple moment system acting on a body can be reduced to a *wrench or screw*: a combination of a *collinear force and couple moment*

4.9 REDUCTION OF A SIMPLE DISTRIBUTED LOADING

In many situations a very large surface area of a body may be subjected to *distributed loadings* such as those caused by wind, fluids or simply the weight of material supported over the body's surface.

- *Distributed loadings* are defined by using a loading function $w = w(x)$ that indicates the intensity of the loading along the length of the member. This intensity is measured in N/m or lb/ft.
- The external effects caused by a coplanar distributed load acting on a body can be represented by a *single resultant force*.
- The *magnitude of the resultant force* is equal to the total *area* under the distributed loading diagram $w = w(x)$.
- The location of the resultant force is given by the fact that it's line of action passes through the *centroid* or geometric center of this area.

HELPFUL TIPS AND SUGGESTIONS

- Be aware of the difference between the *cross product* and the *dot product* of two vectors. The former is a *vector* while the latter is a *scalar*. Also

$$\mathbf{A} \cdot \mathbf{B} = \mathbf{B} \cdot \mathbf{A} \quad \textbf{BUT} \quad \mathbf{A} \times \mathbf{B} \neq \mathbf{B} \times \mathbf{A} \quad (= -(\mathbf{A} \times \mathbf{B}))$$

- The right-hand rule is essential in the calculation of moments. You should be able to apply this rule quickly and accurately.

REVIEW QUESTIONS

1. What is meant by a *moment*? Is it a vector or a scalar?
2. What's the magnitude of the moment of a force about a point?
3. How do you calculate the sense (direction) of a moment?
4. If the line of action of a force passes through a point P, what do you know about the moment of the force about P?
5. If you know the components of two vectors $\mathbf{A}$ and $\mathbf{B}$, how do you determine their cross product? Is the cross-product a vector or a scalar?
6. If $\mathbf{A} \times \mathbf{B} = \mathbf{0}$, what does this mean?
7. When you use the equation $\mathbf{M}_O = \mathbf{r} \times \mathbf{F}$ to determine the moment of a force $\mathbf{F}$ about O, how do you choose $\mathbf{r}$?
8. If you know the components of the vector $\mathbf{M}_O = \mathbf{r} \times \mathbf{F}$, how can you determine the product of the magnitude of $\mathbf{F}$ and the perpendicular distance d from O to the line of action of $\mathbf{F}$?
9. How do you figure out the sense of the moment of $\mathbf{F}$ about O using the formula $\mathbf{M}_O = \mathbf{r} \times \mathbf{F}$?

10. How would you calculate the moment exerted about a point O by a couple consisting of forces $\mathbf{F}$ and $-\mathbf{F}$?

11. True or false? The moment of a couple about O is different than the moment of the same couple about $P \neq O$?

12. What is meant by an *equivalent system*?

13. How do we *replace* a system of forces and couple moments acting on a body with an *equivalent single* resultant force and couple moment acting at a specified point O? What are the relevant equations?

14. Define what is meant by a *wrench*. When does a force and couple moment system acting on a body reduce to a wrench?

15. How is the resultant force exerted by a coplanar distributed load acting on a body determined from the function $w(x)$?

5

Equilibrium of a Rigid Body

MAIN GOALS OF THIS CHAPTER:

- To develop the equations of equilibrium for a rigid body.
- To introduce the concept of the free-body diagram for a rigid body.
- To show how to solve rigid body equilibrium problems using the equations of equilibrium.

5.1 CONDITIONS FOR RIGID-BODY EQUILIBRIUM

A *rigid body* is the next level of sophistication (after the particle) in the modelling of an object. Basically we 'add' size/shape to the existing model of a particle. Consequently, the main difference between a particle and a rigid body is that a rigid body can support moments. To obtain equations for the equilibrium of a rigid body, therefore, we need to supplement the equations of particle equilibrium with an expression of moment balance.

- The *two equations* of equilibrium for a *rigid body* are

$$\sum \mathbf{F} = \mathbf{0},$$
$$\sum \mathbf{M}_O = \mathbf{0},$$

where O is an arbitrary point.

EQUILIBRIUM IN TWO DIMENSIONS

5.2 FREE-BODY DIAGRAMS

- No equilibrium problem should be solved without *first* drawing the free-body diagram, so as to account for *all* the forces *and couple moments* that act on the body.
- Part II of this study pack is devoted to the drawing of free-body diagrams including, specifically, *free-body diagrams for rigid body equilibrium in two dimensions*. Study Part II of this study pack making special note of the following important points:

- If a support *prevents translation* of a body in a particular direction, then the support exerts a *force* on the body in that direction.
- If *rotation is prevented*, then the support exerts a *couple moment* on the body.
- Study Table 2.1 in Part II of this supplement (or Table 5-1 of the text).
- Internal forces are never shown on the free-body diagram since they occur in equal but opposite collinear pairs and therefore cancel out.
- The weight of a body is an external force and its effect is shown as a single resultant force acting through the body's center of gravity G.
- *Couple moments* can be placed anywhere on the free-body diagram since they are *free vectors*. Forces can act at any point along their lines of action since they are *sliding vectors*.

5.3 EQUATIONS OF EQUILIBRIUM

- When the body is subjected to a system of forces which all lie in the x–y plane, the forces can be resolved into their x and y components. Consequently, the conditions for equilibrium in two dimensions can be written in scalar form as:

$$\sum F_x = 0, \quad \sum F_y = 0, \quad \sum M_O = 0, \tag{5.0}$$

where $\sum M_O$ represents the algebraic sum of the couple moments and moments of all the force components about an axis perpendicular to the xy-plane and passing through an arbitrary point O (on or off the body).

TWO ALTERNATIVE SETS OF EQUILIBRIUM EQUATIONS

- $$\sum F_a = 0, \quad \sum M_A = 0, \quad \sum M_B = 0.$$

  Here, the only requirement is that a line passing through points A and B is not perpendicular to the a-axis.

- $$\sum M_A = 0, \quad \sum M_B = 0, \quad \sum M_C = 0.$$

  Here, the only requirement is that points A, B and C do not lie on the same line.

PROCEDURE FOR SOLVING COPLANAR FORCE EQUILIBRIUM PROBLEMS

- **Free-Body Diagram**
  - Establish the x, y coordinate axes in any suitable orientation.
  - Draw an outlined shape of the body.
  - Show all the forces and couple moments acting on the body.
  - Label all the loadings and specify their directions relative to the xy-axes.
  - Indicate the dimensions of the body necessary for computing the moments of forces.
- **Equations of Equilibrium**
  - Apply the moment equation of equilibrium ($\sum M_O = 0$) about a point O that lies at the intersection of the lines of action of two unknown forces. In this way, the moments of these unknowns are zero about O, and a direct solution for the third unknown can be determined.
  - When applying the force equilibrium equations ($\sum F_x = 0$ and $\sum F_y = 0$), orient the x and y axes along lines that will provide the simplest resolution of the forces into their x and y components.
  - If the solution of the equilibrium equations yields a negative scalar for a force or couple moment magnitude, it means that the sense is opposite to that which was assumed on the free-body diagram.

5.4 TWO- AND THREE- FORCE MEMBERS

The solution to some equilibrium problems can be simplified if one is able to recognize members that are subjected to only two or three forces.

- **Two-Force Members**—When a member is subjected to *no couple moments* and forces applied at only two points A and B on a member, the member is called a *two-force member*. In this case, for the member to be in equilibrium, it is necessary that the *resultant* forces at A and B must be *equal, opposite and collinear*. The line of action of both (resultant) forces is known since it always passes through A and B. *Hence only the force magnitude (remember both resultants are equal in magnitude!) needs to be determined or stated.*
- **Three-Force Members**—When a member is subjected to *only three forces*, it is necessary that the forces be *either concurrent or parallel* for the member to be in equilibrium. Once the point of concurrency O (where the lines of action of the forces intersect) is identified, then necessarily $\sum M_O = 0$. If two of the three forces are parallel, the point of concurrency O, is said to be at "infinity" and the third force must be parallel to the other two forces to intersect at this "point."

EQUILIBRIUM IN THREE DIMENSIONS

5.5 FREE-BODY DIAGRAMS

The first step in solving three-dimensional equilibrium problems, as in the case of two dimensions, is to *draw a free-body diagram.* The general procedure for doing this is the same as that outlined for the two-dimensional case in Section 5.2 of the text. However, there are a few subtle differences of which you should be aware:

- It is necessary to be familiar with the different types of reactive forces and couple moments acting at various types of supports and connections when members are viewed in three dimensions. It is important to recognize the symbols used to represent each of these supports and to understand clearly how the forces and couple moments are developed by each support. These are summarized in Table 5-2 of the text. Remember:
  - *As in the two-dimensional case, a force is developed by a support that restricts the translation of the attached member, whereas a couple moment is developed when rotation of the attached member is prevented.*

5.6 EQUATIONS OF EQUILIBRIUM

When the body is subjected to a three-dimensional force system, equilibrium requires that the resultant force and resultant couple moment acting on the body be equal to zero.

- In *vector* form the *two* equilibrium equations are

$$\sum \mathbf{F} = \mathbf{0},$$
$$\sum \mathbf{M}_O = \mathbf{0},$$

  where $\sum \mathbf{F}$ is the vector sum of all the external forces acting on the body and $\sum \mathbf{M}_O$ is the sum of the couple moments and the moments of all the forces about any point O (on or off the body).
- Writing

$$\sum \mathbf{F} = \sum F_x \mathbf{i} + \sum F_y \mathbf{j} + \sum F_z \mathbf{k},$$
$$\sum \mathbf{M}_O = \sum M_x \mathbf{i} + \sum M_y \mathbf{j} + \sum M_z \mathbf{k},$$

  The *six scalar* equilibrium equations are

$$\sum F_x = 0, \quad \sum F_y = 0, \quad \sum F_z = 0, \qquad (5.1)$$
$$\sum M_x = 0, \quad \sum M_y = 0, \quad \sum M_z = 0.$$

5.7 CONSTRAINTS AND STATISTICAL DETERMINANCY

To ensure equilibrium of a rigid body, it is not only necessary to satisfy the equations of equilibrium, but the body must also be properly held or constrained by its supports.

- **Redundant Constraints.** When a body has *redundant supports*, that is, more supports than are necessary to hold it in equilibrium, it becomes *statically indeterminate*. This means that there will be more unknown loadings on the body than equations of equilibrium available for their solution. The additional equations needed to solve indeterminate problems are generally obtained from the *deformation conditions* at the points of support. These equations involve modelling the body not as a rigid body but as a *deformable body* (the next level of sophistication). This is done in courses dealing with "mechanics of materials."
- **Improper Constraints**. In some cases, there may be as many unknown forces on the body as there are equations of equilibrium; however, *instability* of the body may develop because of improper constraining by the supports. When this happens, either the number of available equilibrium equations is reduced by one (making the system *indeterminate*) or we will not be able to satisfy *all* the equilibrium equations. Proper constraining (avoiding instability of a body) requires

  1. The lines of action of the reactive forces do not intersect a common axis **and**
  2. The reactive forces must not all be parallel to one another.

When the minimum number of reactive forces is needed to properly constrain the body in question, the problem will be *statically determinate* and therefore the equations of equilibrium can be used to determine *all* the reactive forces.

HELPFUL TIPS AND SUGGESTIONS

- The first step in solving equilibrium problems is to draw a *free-body diagram*. Don't try to skip this stage no matter how trivial you think it is!
- Make the *free-body diagram* as clear and concise as possible. It will aid your understanding of the problem and it will help you construct the equilibrium equations.

REVIEW QUESTIONS

1. Why is there no moment equilibrium equation for a body modelled as a particle?
2. Write down the six independent scalar equilibrium equations for a rigid body in three dimensions. Adapt these equations to the two-dimensional case explaining why there are now only three independent equations.
3. What does it mean when an object is said to have redundant supports.
4. How do you know if an object is statically indeterminate as a result of redundant supports?
5. How do you avoid instability of a body due to improper constraining?

6

Structural Analysis

MAIN GOALS OF THIS CHAPTER:

- To show how to determine the forces in the members of a truss using the method of joints and the method of sections.
- To analyze the forces acting on the members of frames and machines composed of pin-connected members.

6.1 SIMPLE TRUSSES

A *truss* is a structure composed of slender members joined together at their end points. The members are usually wooden struts or metal bars. The joint connections are usually formed by bolting or welding the ends of the members to a common plate called a *gusset plate*, or by simply passing a large bolt or pin through each of the members.

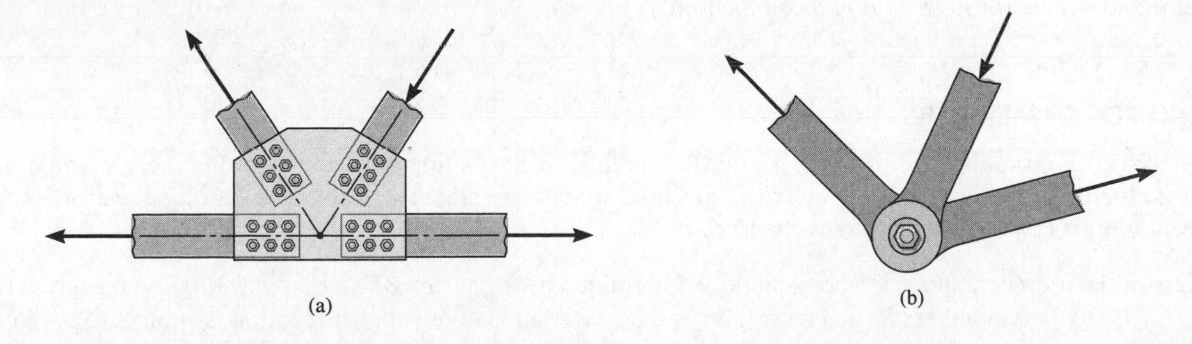

(a) (b)

- **Planar Trusses** lie in a single plane and are often used to support roofs and bridges.
- **Assumptions for Design.** To design both the members and the connections of a truss, it is first necessary to determine the force developed in each member when the truss is subjected to a given loading. The following assumptions allow us to consider each truss member as a *two-force member* so that the forces at the ends of the member must be directed along the axis of the member.
  - *All loadings are applied at the joints.*
  - *The members are joined together by smooth pins.*

- **Simple Trusses.** To prevent collapse, the form of a truss must be rigid. The simplest form which is rigid or stable is a *triangle.* Consequently, a *simple truss* is constructed by *starting* with a basic triangular element. Additional elements consisting of two members and a joint are added to the triangular element to form a *simple truss*.

6.2 THE METHOD OF JOINTS

In order to analyze or design a truss, we must obtain the force in each of its members. To do this, we consider the *equilibrium of a joint* of the truss. This is the basis for the *method of joints*.

- Since the truss members are all straight two-force members lying in the same plane, the force system acting at each joint is *coplanar and concurrent.* Consequently, moment equilibrium is automatically satisfied at the joint and it is only necessary to satisfy *two independent scalar force equilibrium equations.*

PROCEDURE FOR ANALYZING A (PLANAR) TRUSS USING THE METHOD OF JOINTS

- Draw the free-body diagram of a joint having at least one known force and at most two unknown forces. (If this joint is at one of the supports, it generally will be necessary to know the external reactions at the truss support).
- Establish the sense of an unknown force by either:
  - Assuming that the unknown force is in *tension* and interpreting negative scalar results as members in *compression*. OR
  - *By inspection.*
- Orient the x and y axes such that the forces on the free-body diagram can be easily resolved into their x and y components and then apply the two force equilibrium equations $\sum F_x = 0$ and $\sum F_y = 0$. Solve for the two unknown member forces and verify their correct sense.
- Continue to analyze each of the other joints as above.
- Once the force in a member is found from the analysis of a joint at one of its ends, the result can be used to analyze the forces acting on the joint at its other end. Remember that a member in *compression pushes* on the joint and a member in *tension pulls* on the joint.

6.3 ZERO FORCE MEMBERS

Truss analysis using the method of joints is greatly simplified if one is first able to determine those members which support no loading. These *zero-force members* are used to increase stability of the truss during construction and to provide support if the applied loading is changed.

- Zero-force members of a truss are generally determined by *inspection of each of its joints.* As a general rule:

  If only two members form a truss joint and no external load or support reaction is applied to the joint, the members must be zero force members.

  If three members form a truss joint for which two of the members are collinear, the third member is a zero force member provided no external force or support reaction is applied to the joint.

6.4 THE METHOD OF SECTIONS

The *method of sections* is used to determine the loadings acting within a body. It is based on the principle that *if a body is in equilibrium then any part (section) of the body is also in equilibrium.*

PROCEDURE FOR ANALYZING THE FORCES IN THE MEMBERS OF A TRUSS USING THE METHOD OF SECTIONS

- **Free-Body Diagram**
  - Make a decision as to how to "cut" or section the truss through the members where forces are to be determined.
  - Before isolating the appropriate section, it may first be necessary to determine the truss' *external* reactions. Then three equilibrium equations are available to solve for member forces at the cut section
  - Draw the free-body diagram of that part of the sectioned truss which has the least number of forces acting on it.
  - Establish the sense of an unknown member force by either:
    - ∗ Assuming that the unknown member force is in *tension* and interpreting negative scalar results as members in *compression*. OR
    - ∗ *By inspection.*
- **Equations of Equilibrium**
  - Moments should be summed about a point that lies at the intersection of the lines of action of two unknown forces, so that the third unknown force is determined directly from the moment equation.
  - If two of the unknown forces are *parallel*, forces may be summed *perpendicular* to the direction of these unknowns to determine *directly* the third unknown force.

6.5 SPACE TRUSSES

A *space truss* consists of members joined together at their ends to form a stable three-dimensional structure. The simplest element of a space truss is a *tetrahedron*, formed by connecting six members together.

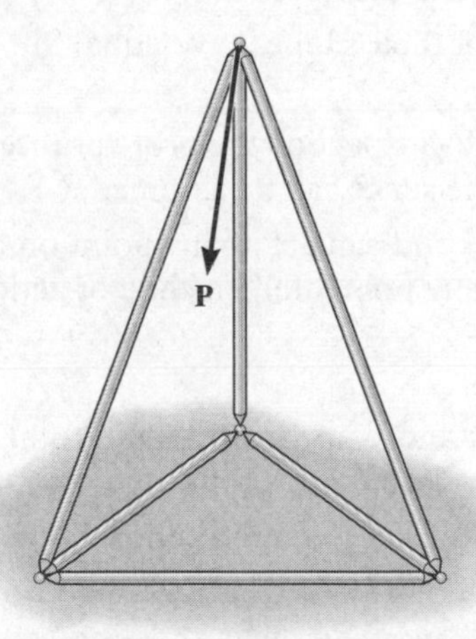

Either the *method of joints* or the *method of sections* can be used to determine the forces developed in the members of a simple space truss:

- **Method of Joints**. If the forces in *all* the members of the truss must be determined, the method of joints is most suitable for the analysis. Solve the three scalar equilibrium equations $\sum F_x = 0$, $\sum F_y = 0$ and $\sum F_z = 0$ at each joint. The solution of many simultaneous equations can be avoided if the force analysis begins at a joint having at least one known force and at most three unknown forces. Use a Cartesian vector analysis if the three-dimensional geometry of the force-system at the joint is hard to visualize.

- **Method of Sections**. If only a *few* member forces are to be determined, the method of sections is most suitable. When an imaginary section is passed through a truss and the truss is separated into two parts, the force system acting on one of the parts must satisfy the six scalar equilibrium equations $\sum F_x = 0$, $\sum F_y = 0$, $\sum F_z = 0$, $\sum M_x = 0$ and $\sum M_y = 0$ and $\sum M_z = 0$. By proper choice of the section and axes for summing forces and moments, many of the unknown member forces in a space truss can be computed *directly* using a single equilibrium equation.

6.6 FRAMES AND MACHINES

Frames and machines are two common types of structures which are often composed of pin-connected *multiforce* members. *Frames* are generally stationary and are used to support loads while *machines* contain moving parts and are designed to transmit and alter the effect of forces. Once the forces at the joints are obtained (see below) it is then possible to *design* the size of the members, connections and supports using the theory of mechanics of materials (deformable bodies) and an appropriate engineering design code.

PROCEDURE FOR DETERMINING THE JOINT REACTIONS ON FRAMES OR MACHINES COMPOSED OF MULTIFORCE MEMBERS

- **Free-Body Diagram**
  - Draw the free-body diagram of the entire structure, a portion of the structure, or each of its members. The choice should be made so that it leads to the most direct solution of the problem.
  - When the free-body diagram of a group of members of a structure is drawn, the forces at the connected parts of this group are *internal* forces and are not shown on the free-body diagram of the group.
  - Forces common to two members which are in contact act with equal magnitude but opposite sense on the respective free-body diagrams of the members.
  - Two-force members, regardless of their shape, have equal but opposite collinear forces acting at the ends of the member.
  - In many cases it is possible to tell by inspection the proper sense of the unknown forces acting on a member; however, if this seems difficult, the sense can be assumed.
  - A couple moment is a free vector and can act at any point on the free-body diagram. Also, a force vector is a sliding vector and can act at any point along its line of action.
- **Equations of Equilibrium**
  - Count the number of unknowns and compare it to the total number of equilibrium equations that are available, In two dimensions, there are three equilibrium equations that can be written for each member.
  - Sum moments about a point that lies at the intersection of the lines of action of as many unknown forces as possible
  - If the solution of a force or couple moment magnitude is negative, it means that the sense is the reverse of that shown on the free-body diagrams.

HELPFUL TIPS AND SUGGESTIONS

- The importance of drawing and using a clear and concise free-body diagram cannot be overstated.
- As in most mechanics problems, *practice* is the key. Make sure you read Examples 6-9 through 6-13 in the text and attempt to draw the requested free-body diagrams *yourself.* When doing so, make sure the work is neat and that all the forces and couple moments are properly labelled.

REVIEW QUESTIONS

1. What is a truss?
2. What assumptions allow us to consider a truss member as a *two-force member?*
3. What is the method of joints?
4. How many independent scalar equilibrium equations are available from the free-body diagram of a joint?
5. What is the method of sections?
6. What methods are available to determine the forces developed in the members of a simple space truss?
7. What's the difference between a frame and a machine?
8. When the free-body diagram of a group of members of a structure is drawn, the forces at the connected parts of this group are not shown on the free-body diagram of the group. Why?

7

Internal Forces

MAIN GOALS OF THIS CHAPTER:

- To show how to use the method of sections for determining the internal loadings in a member.
- To generalize this procedure by formulating equations that can be plotted so that they describe the internal shear and moment throughout a member.
- To analyze the forces and study the geometry of cables supporting a load.

7.1 INTERNAL FORCES DEVELOPED IN STRUCTURAL MEMBERS

The design of any structural or mechanical member requires an investigation of **both** the external loads and reactions acting on the member **and** the loading acting *within* the member—in order to be sure *the material can resist this loading*. These internal loadings can be determined using the *method of sections*.

> The idea is to cut an 'imaginary section' through the member so that the internal loadings (of interest) at the section become external on the free-body diagram of the section.

PROCEDURE FOR FINDING THE INTERNAL LOADINGS AT A SPECIFIC LOCATION IN A MEMBER USING THE METHOD OF SECTIONS

- **Support Reactions.**
  - Before the member is "cut" or "sectioned," it may first be necessary to determine the member's support reactions, so that the equilibrium equations are used only to solve for the internal loadings when the member is sectioned.
  - If the member is part of a *frame or machine*, the reactions at its connections are determined using the methods outlined in Section 6.6.

- **Free-Body Diagram.**
  - Keep all distributed loadings, couple moments and forces acting on the member in their *exact locations*, then pass an imaginary section through the member, perpendicular to its axis at the point where the internal loading is to be determined.
  - After the section is made, draw a free-body diagram of the segment that has the least number of loads on it, and indicate the x, y, z components of the force and couple moment resultants at the section.
  - If the member is subjected to a *coplanar system of forces*, only **N** (normal force), **V** (shear force), and **M** (bending moment) act at the section.

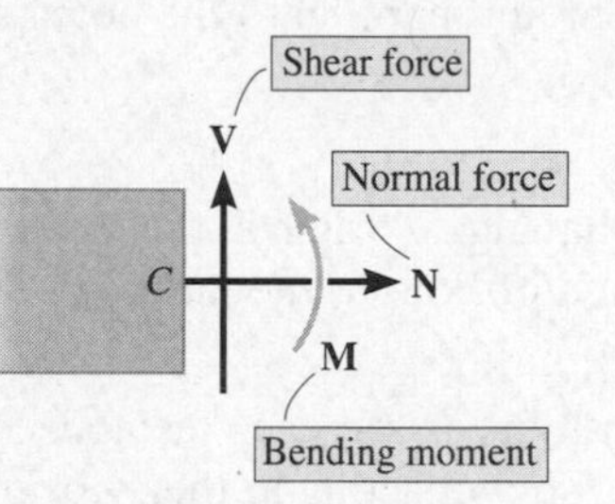

  - In three dimensions, a general internal force and couple moment resultant will act at the section.

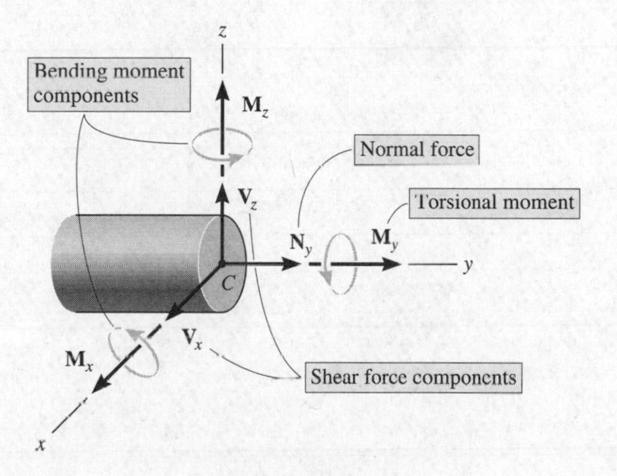

  - In many cases it may be possible to tell by inspection the proper sense of the unknown loadings; however, if this seems difficult, the sense can be assumed.
- **Equations of Equilibrium.**
  - Moments should be summed at the section about axes passing through the *centroid* or geometric center of the member's cross-sectional area in order to eliminate the unknown normal and shear forces and thereby obtain direct solutions for the moment components.
  - If the solutions of the equilibrium equations yields a negative scalar, the assumed sense of the quantity is opposite to that shown on the free-body diagram.

7.2 SHEAR AND MOMENT EQUATIONS AND DIAGRAMS

Beams are designed to support loads *perpendicular* to their axes. The actual design of a *beam* requires a detailed knowledge of the *variation* of the internal shear force V and bending moment M acting at *each point* along the axis of the beam. After this, the theory of mechanics of materials is used with an appropriate engineering design code to determine the beam's required cross-sectional area.

The *variations* of V and M as functions of the position x along the beam's axis can be obtained using the method of sections (Section 7.1). However, it is necessary to section the beam at an arbitrary distance x from one end rather than at a specified point. If the results are plotted, the graphical variations of V and M as functions of x are termed the *shear diagram* and *bending moment diagram*, respectively.

These diagrams can be constructed as follows:

- **Support Reactions.**
  - ◆ Determine all the reactive forces and couple moments acting on the beam and resolve all the forces into components acting perpendicular and parallel to the beam's axis.
- **Shear and Moment Functions.**
  - ◆ Specify separate coordinates x having an origin at the *beam's left end* and extending to regions of the beam *between* concentrated forces and/or couple moments, or where there is no discontinuity of distributed loading.
  - ◆ Section the beam perpendicular to its axis at each distance x and draw the free-body diagram of one of the segments. Be sure **V** and **M** are shown acting in their *positive sense* in accordance with the following sign convention:

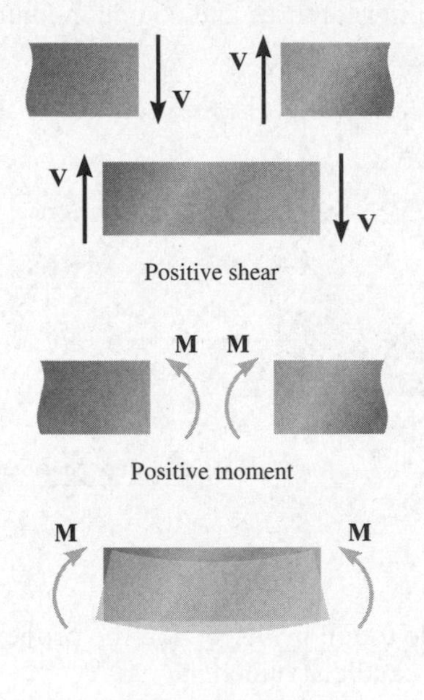

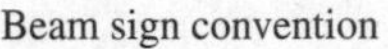

Beam sign convention

  - ◆ The shear V is obtained by summing forces perpendicular to the beam's axis.
  - ◆ The moment M is obtained by summing moments about the sectioned end of the segment.
- **Shear and Moment Diagrams.**
  - ◆ Plot the shear diagram (V versus x) and the moment diagram (M versus x). If the computed values of the functions describing V and M are *positive*, the values are plotted above the x-axis, whereas *negative* values are plotted below the x-axis.
  - ◆ Generally, it is convenient to plot the shear and bending-moment diagrams directly below the free-body diagram of the beam. See Examples 7-7 and 7-8 in the text.

7.3 RELATIONS BETWEEN DISTRIBUTED LOAD, SHEAR AND MOMENT

In cases where a beam is subjected to several concentrated forces, couple moments, and distributed loads, the method of constructing the shear and bending moment diagrams may become quite tedious. In this section a simpler method for constructing these diagrams is presented—based on differential relations that exist between the load, shear, and bending moment. The following are the main points:

- The slope of the shear diagram is equal to the negative of the intensity of the distributed loading, where positive distributed loading is downward i.e.

$$\frac{dV}{dx} = -w(x). \tag{7.0}$$

- If a concentrated force acts downward on the beam, the shear will jump downward by the amount of the force.
- The change in the shear ΔV between two points is equal to the *negative of the area* under the distributed-loading curve between the points.
- The slope of the moment diagram is equal to the shear i.e.

$$\frac{dM}{dx} = V. \tag{7.1}$$

- The change in the moment ΔM between two points is equal to the *area* under the shear diagram between the two points.
- If a *clockwise* couple moment acts on the beam, the shear will not be affected, however, the moment diagram will jump *upward* by the amount of the moment.
- Points of zero shear represent points of *maximum or minimum moment* since

$$\frac{dM}{dx} = 0. \tag{7.2}$$

7.4 CABLES

Flexible cables and chains are used in engineering structures for support and to transmit loads from one member to another. In the force analysis of such systems, the weight of the cable itself may be neglected (cable is referred to as '*weightless*') because it is often small compared to the load it carries. In modelling the cable, it is assumed that:

1. The cable is *perfectly flexible* (cable offers no resistance to bending so the tensile force acting in the cable is always *tangent* to the cable at points along its length).
2. The cable is *inextensible* (cable has a constant length before and after load is applied—cable can be treated as a rigid body).

CABLE SUBJECTED TO CONCENTRATED LOADS

- When a cable of negligible weight supports several concentrated loads, the cable takes the form of several straight-line segments, each of which is subjected to a constant tensile force. The equilibrium analysis is performed by writing down a *sufficient number* of equilibrium equations (based on the entire cable or any part thereof) and equations describing the geometry of the cable to solve for all the *unknowns* leading to a description of the tension in (each segment of) the cable. See Example 7-13 in text.

CABLE SUBJECTED TO A DISTRIBUTED LOAD

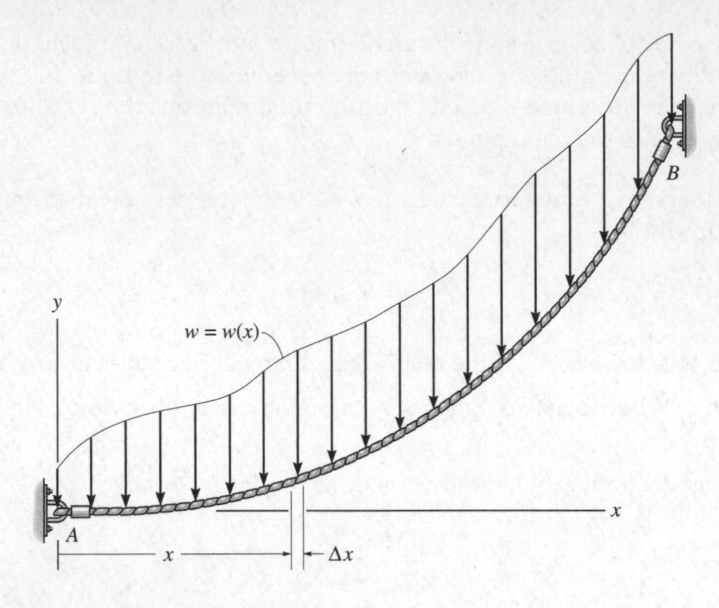

- The equation

$$y = \frac{1}{F_H} \int \left(\int w(x)\, dx \right) dx, \tag{7.3}$$

determines the curve for the cable $y = f(x)$. Here, F_H measures the horizontal component of tensile force at *any point* along the cable and $w(x)$ is the loading function measured in the x-direction. In practice, F_H and the two constants of integration are determined from the boundary conditions for the cable—see Example 7-14 in text.

CABLE SUBJECTED TO ITS OWN WEIGHT

- When the weight of the cable becomes important in the force analysis (e.g., cables used as transmission lines), the loading function along the cable becomes a function of the arc length s rather than the projected length x. The equation of the deflection curve is given by $y = f(x)$ where

$$\frac{dy}{dx} = \frac{1}{F_H} \int w(s)\, ds, \tag{7.4}$$

$$x = \int \frac{ds}{\left\{1 + \frac{1}{F_H^2} \left(\int w(s)\, ds\right)^2\right\}^{\frac{1}{2}}}. \tag{7.5}$$

The two constants of integration from (7.5) are found using the boundary conditions for the cable. First solve (7.5) for $w(s)$ then use (7.4) to get the shape $y = f(x)$ for the cable. See Example 7-15 in text.

HELPFUL TIPS AND SUGGESTIONS

- Use *clear* and concise free-body diagrams.
- As in most mechanics problems, *practice* is the key. Make sure you read Examples 7-1 through 7-15 in the text before you attempt any corresponding problems. These examples will serve as templates with which to solve problems. Draw the requested free-body diagrams *yourself.* When doing so, make sure the work is neat and that all the forces and couple moments are properly labelled.

REVIEW QUESTIONS

1. What are the normal (or axial) force, shear force and bending moment?
2. How are the positive directions (sense) of the shear force **V** and bending moment **M** defined?
3. For a portion of a beam which is subjected only to a distributed load w, how are the shear force and bending moment distributions determined from equations (7.0) and (7.1) ?
4. What does it mean when a cable is assumed to be 'weightless,' 'inextensible' or 'perfectly flexible'?
5. If a cable is subjected to a load that is uniformly distributed along a straight line and its weight is negligible, what mathematical curve describes its shape?

8

Friction

MAIN GOALS OF THIS CHAPTER:

- To introduce the concept of dry friction and show how to analyze the equilibrium of rigid bodies subjected to this force.
- To present specific applications of frictional force analysis on wedges, screws, belts and bearings.
- To investigate the concept of *rolling resistance*.

8.1 CHARACTERISTICS OF DRY FRICTION

As a result of *experiments* that pertain to the foregoing discussion, the following rules which apply to bodies subjected to dry friction may be stated:

- The frictional force acts *tangent* to the contacting surfaces in a direction *opposed to the relative motion* or tendency for motion of one surface against another.
- The maximum static frictional force $\mathbf{F}_s$ that can be developed is independent of the area of contact, provided the normal pressure is not very low nor great enough to severely deform or crush the contacting surfaces of the bodies.
- The maximum static frictional force is generally greater than the kinetic frictional force $\mathbf{F}_k$ for any two surfaces of contact. However, if one of the bodies is moving with a *very low velocity* over the surface of another, F_k becomes approximately equal to F_s i.e. $\mu_s \approx \mu_k$.
- When *slipping* at the surface of contact is *impending (about to occur)*, the maximum static frictional force is proportional to the normal force $\mathbf{N}$, such that $F_s = \mu_s N$.
- When *slipping* at the surface of contact is *occurring*, the kinetic frictional force is proportional to the normal force $\mathbf{N}$, such that $F_k = \mu_k N$.

8.2 PROBLEMS INVOLVING DRY FRICTION

If a rigid body is in equilibrium when it is subjected to a system of forces which includes the effect of friction, the force system must satisfy not only the equations of equilibrium but *also* the laws that govern the frictional forces.

In general, there are three types of mechanics problems involving dry friction. They are classified once the free-body diagrams are drawn and the total number of unknowns are identified and compared with the total number of available equilibrium equations:

- **Equilibrium**—The total number of unknowns is equal to the total number of available equilibrium equations. In this case, once the frictional forces are determined, check that $F \leq \mu_s N$ otherwise slipping will occur and the body will not remain in equilibrium.
- **Impending Motion at all Points**—The total number of unknowns will *equal* the total number of available equilibrium equations plus the total number of available frictional equations or conditional equations for tipping. As a result, several possibilities for motion or impending motion will exist and the problem will involve a determination of the kind of motion which actually occurs.
- **Impending Motion at Some Points**—The total number of unknowns will be *less* than the total number of available equilibrium equations plus the total number of available frictional equations, $F = \mu N$. If motion is *impending* at the points of contact, then $F_s = \mu_s N$ whereas if the body is *slipping*, then $F_k = \mu_k N$.

EQUILIBRIUM VERSUS FRICTIONAL EQUATIONS

When the frictional force **F** is an equilibrium force i.e., $F < \mu N$, we can always *assume* the sense of the frictional force **F** (since the frictional force *always* acts so as to oppose the relative motion or impede the motion of a body over the contacting surface). The correct sense is determined after solving the equilibrium equations for F. However, in cases where $F = \mu N$ is used, we can no longer assume the sense of **F** since the equation $F = \mu N$ relates only the magnitudes of two perpendicular vectors. Consequently, in this case, **F** must always be shown acting with its correct sense on the free-body diagram.

PROCEDURE FOR ANALYSIS

- **Free-Body Diagrams**
  - Draw the necessary free-body diagrams and, unless it is stated in the problem that impending motion or slipping occurs, *always* show the frictional forces as unknowns i.e., *do not assume* $F = \mu N$.
  - Determine the number of unknowns and compare this with the number of available equilibrium equations.
  - If there are more unknowns than equilibrium equations, it is necessary to apply the frictional equations at some, if not all, points of contact to obtain the extra equations needed for complete solution.
  - If the equation $F = \mu N$ is to be used, it will be necessary to show **F** acting in the proper direction on a free-body diagram.
- **Equations of Equilibrium and Friction.**
  - Apply the equations of equilibrium and the necessary frictional equations (or conditional equations if tipping is possible) and solve for the unknowns.
  - If the problem involves a three-dimensional force system such that it becomes difficult to obtain the force components or the necessary moment arms, apply the *vector equations of equilibrium*.

8.3 WEDGES

A *wedge* is a simple machine which is often used to transform an applied force into much larger forces, directed at approximately right angles to the applied force. Also wedges can be used to give small displacements or adjustments to heavy loads. The analysis of problems involving wedges proceeds as above i.e., we draw free-body diagrams of the wedge and any other contacting bodies and formulate the appropriate equilibrium and frictional equations. See Example 8-7 in text.

8.4 FRICTIONAL FORCES ON SCREWS

A *screw* may be thought of simply as an inclined plane or wedge wrapped around a cylinder. In most cases screws are used as fasteners; however, in many applications, they are incorporated to transmit power or motion from one part of the machine to another.

Before proceeding to solve problems involving frictional forces on screws, each of the following cases should be thoroughly understood:

- **Frictional Analysis with Upward Screw Motion.** The moment necessary to cause upward impending motion of the screw is

$$M = Wr\tan(\theta + \phi), \quad \phi = \phi_s = \tan^{-1}\mu_s. \tag{8.0}$$

If ϕ is replaced by $\phi_k = \tan^{-1}\mu_k$, we obtain a smaller value of M necessary to maintain uniform upward motion of the screw.

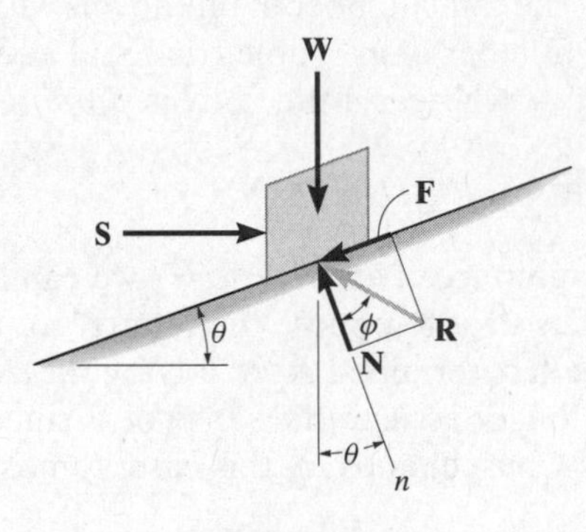

Upward screw motion

- **Frictional Analysis with Downward Screw Motion** (when the surface of the screw is very slippery: $\theta > \phi$). The moment necessary to cause downward impending motion of the screw is

$$M' = Wr\tan(\theta - \phi), \quad \phi = \phi_s. \tag{8.1}$$

If ϕ is replaced by $\phi_k = \tan^{-1}\mu_k$, we obtain a (smaller) value of M necessary to maintain *uniform* downward motion of the screw.

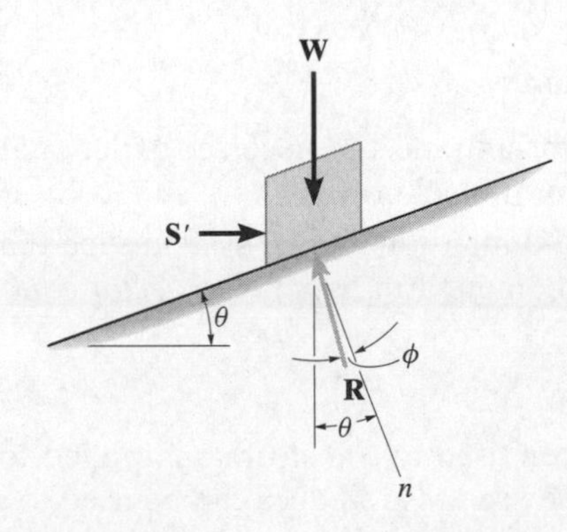

Downward screw motion ($\theta > \phi$)

- **Frictional Analysis with Downward Screw Motion** (when the surface of the screw is very rough: $\theta < \phi$). The moment necessary to cause downward impending motion of the screw is

$$M' = Wr\tan(\phi - \theta), \quad \phi = \phi_s. \tag{8.2}$$

If ϕ is replaced by $\phi_k = \tan^{-1}\mu_k$, we obtain a (smaller) value of M necessary to maintain uniform downward motion of the screw.

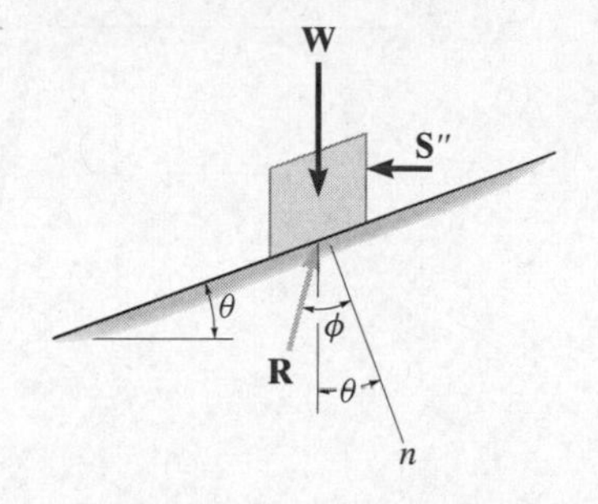

Downward screw motion ($\theta < \phi$)

- **Frictional Analysis with a Self-Locking Screw**. If the moment M (or its effect S) is removed, the screw will remain *self-locking* i.e. it will support the load W by friction forces alone (provided $\phi \geq \theta$).

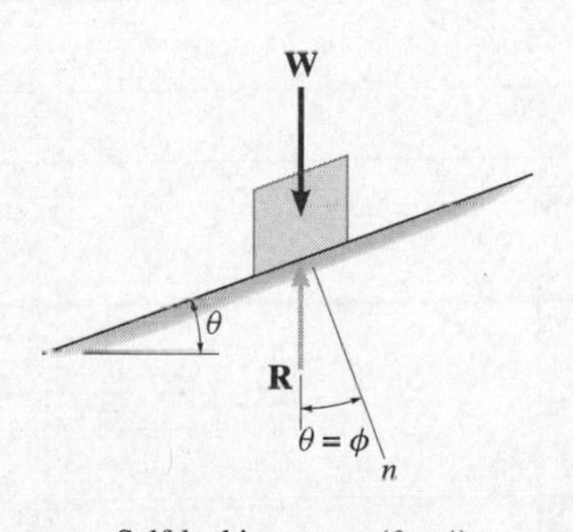

Self-locking screw ($\theta = \phi$)
(on the verge of rotating downward)

8.5 FRICTIONAL FORCES ON FLAT BELTS

Whenever belt drives are designed, it is necessary to dtermine the frictional forces developed between the belt and its contacting surface.

- The tension T_2 in the belt required to pull the belt counterclockwise over the surface and thereby overcome both the frictional forces at the surface of contact and the known tension T_1 (motion or impending motion of belt relative to surface) is:

$$T_2 = T_1 e^{\mu\beta}$$

where μ is the coefficient of static or kinetic friction between the belt and the surface of contact, and β is in radians.

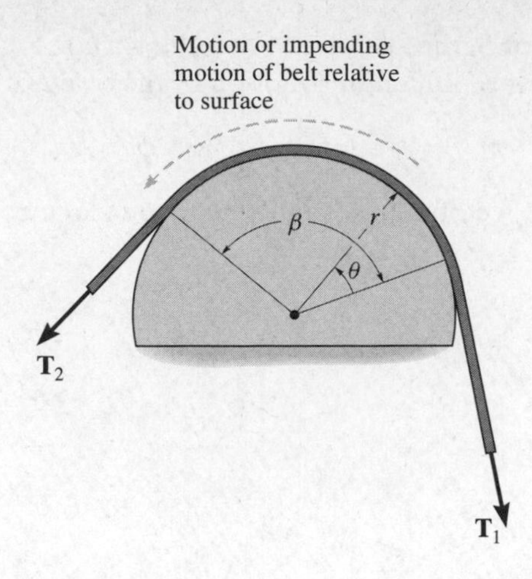

8.6 FRICTIONAL FORCES ON COLLAR BEARINGS, PIVOT BEARINGS, AND DISKS

Pivot and collar bearings are commonly used in machines to support an axial load on a rotating shaft.

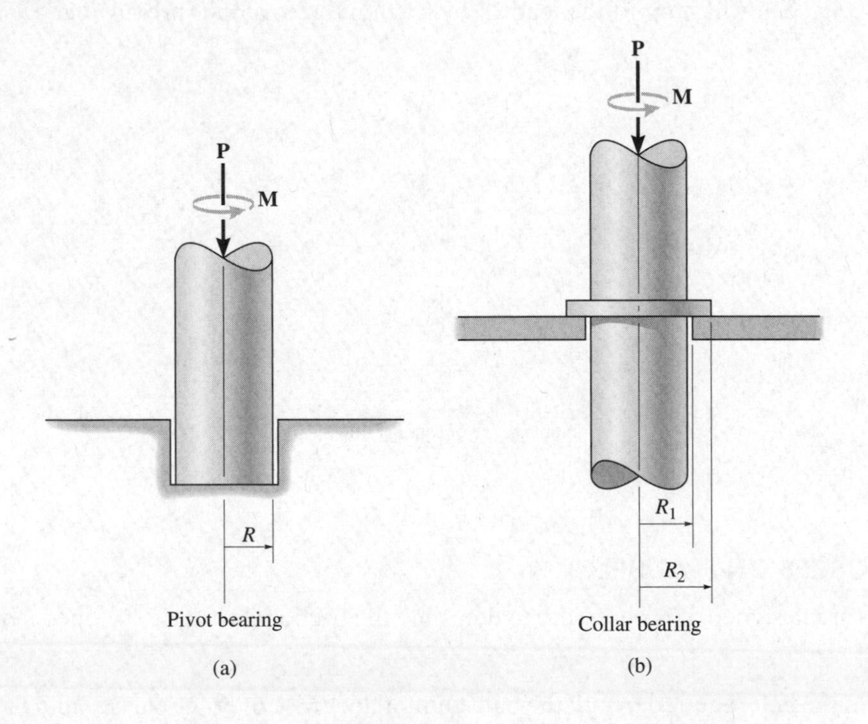

- The magnitude of the moment required for impending rotation of the shaft is given by

$$M = \frac{2}{3}\mu_s P \left(\frac{R_2^3 - R_1^3}{R_2^2 - R_1^2} \right).$$

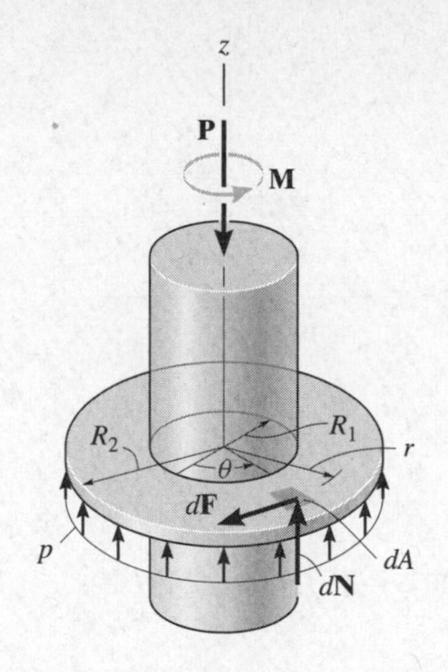

- The frictional moment developed at the end of the shaft, when it is rotating at constant speed can be found by substituting μ_k for μ_s in the expression for M.

8.7 FRICTIONAL FORCES ON JOURNAL BEARINGS

When a shaft or axle is subjected to lateral loads, a *journal bearing* is commonly used for support.

- The moment needed to maintain constant rotation of the shaft is given by

$$M = Rr\sin\phi_k, \tag{8.3}$$

where ϕ_k is the angle of kinetic friction defined by $\tan\phi_k = \mu_k$ and R is the magnitude of the bearing reactive force acting at A.

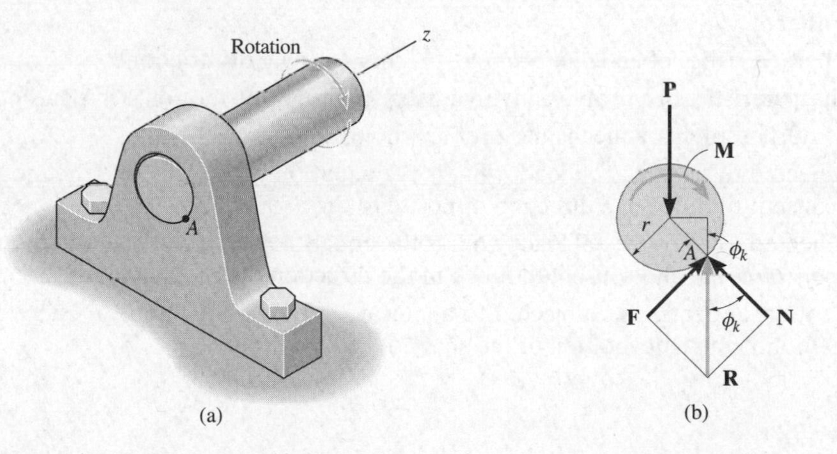

8.8 ROLLING RESISTANCE

- The force **P** necessary to initiate and maintain rolling (of a cylinder with weight **W** and radius r) at constant velocity has magnitude

$$P \approx \frac{Wa}{r}.$$

Here, the *distance a* is referred to as the *coefficient of rolling resistance*.

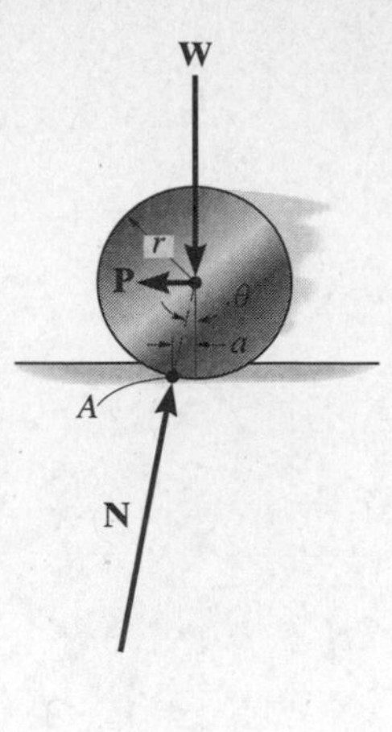

HELPFUL TIPS AND SUGGESTIONS

- Use *clear* and concise free-body diagrams.
- As in most mechanics problems, *practice* is the key. Make sure you read Examples 7-1 through 7-15 in the text before you attempt any corresponding problems. These examples will serve as templates with which to solve problems. Draw the requested free-body diagrams *yourself.* When doing so, make sure the work is neat and that all the forces and couple moments are properly labelled.

REVIEW QUESTIONS

1. If relative slipping of two dry surfaces in contact is impending what can you say about the frictional forces they exert on each other?
2. If two dry surfaces in contact are sliding relative to each other, what can you say about the frictional forces they exert on each other?
3. What are the characteristics of an *equilibrium* problem involving dry friction?
4. What are the characteristics of a problem involving dry friction when motion is impending at all points of contact?
5. What's the first step in solving a mechanics problem involving dry friction?
6. If a screw is subjected to a large axial load, what is the equation which will give the moment necessary to rotate the screw and cause it to move in a direction opposite to the axial load?
7. If a screw is subjected to a large axial load, what is the equation which will give the moment necessary to rotate the screw at *a constant rate* and cause it to move in the direction of the axial load?
8. If the shaft of a journal bearing is subjected to a lateral load with magnitude P, how do you find the moment necessary to maintain constant rotation of the shaft?

9

Center of Gravity and Centroid

MAIN GOALS OF THIS CHAPTER:

- To discuss the concept of the *center of gravity, center of mass*, and the *centroid*.
- To show how to determine the location of the center of gravity and centroid for a system of discrete particles and a body of arbitrary shape.
- To use the theorems of Pappus and Guldinus for finding the area and volume for a surface of revolution.
- To present a method for finding the resultant of a general distributed loading and show how it applies to finding the resultant of a fluid.

9.1 CENTER OF GRAVITY, CENTER OF MASS AND THE CENTROID OF A BODY

- The *center of gravity* G is a point which locates the *resultant weight of a system of particles*. This coordinates of the center of gravity G of a system of particles is given by

$$\bar{x} = \frac{\sum \tilde{x} W}{\sum W}, \quad \bar{y} = \frac{\sum \tilde{y} W}{\sum W}, \quad \bar{z} = \frac{\sum \tilde{z} W}{\sum W}, \tag{9.0}$$

  where $\tilde{x}$, $\tilde{y}$, $\tilde{z}$ represent the coordinates of each particle of the system and $\sum W$ is the resultant sum of the weights of all the particles in the system.
- The *center of mass* of a system of particles is obtained by substituting $W = mg$ (assuming g for every particle is constant) into (9.0). Consequently, the center of mass has coordinates

$$\bar{x} = \frac{\sum \tilde{x} m}{\sum m}, \quad \bar{y} = \frac{\sum \tilde{y} m}{\sum m}, \quad \bar{z} = \frac{\sum \tilde{z} m}{\sum m},$$

  where $\sum W$ is the resultant sum of the masses of all the particles in the system. Note that the *location* of the center of gravity *coincides* with that of the center of mass. However, the *center of mass is independent of gravity* and so can be used in situations when particles are not under the influence of a gravitational attraction.

- The coordinates of the *center of gravity* G of a body are given by

$$\bar{x} = \frac{\int \tilde{x}\, dW}{\int dW}, \quad \bar{y} = \frac{\int \tilde{y}\, dW}{\int dW}, \quad \bar{z} = \frac{\int \tilde{z}\, dW}{\int dW}, \tag{9.1}$$

where $\tilde{x}, \tilde{y}, \tilde{z}$ represent the coordinates of an arbitrary point in the body.

- The *center of mass* of a rigid body is obtained by substituting $dW = g$ into (9.1) and cancelling g from both the numerators and denominators. This yields (9.1) with W replaced by m.

CENTROID

- The *centroid* $C\ (\bar{x}, \bar{y}, \bar{z})$ is a point which defines the *geometric center of a body*. This point coincides with the *center of mass* or the *center of gravity* only if the material composing the body is *uniform or homogeneous* (in which case both γ and ρ are constant throughout the body).
- Formulas used to *locate* the center of gravity or the centroid simply represent a balance between the sum of moments of all the parts of the system and the moment of the "resultant" for the system. There are three cases to consider:
  - **Volume**. $C\ (\bar{x}, \bar{y}, \bar{z})$ is given by

$$\bar{x} = \frac{\int_V \tilde{x}\, dV}{\int_V dV}, \quad \bar{y} = \frac{\int_V \tilde{y}\, dV}{\int_V dV}, \quad \bar{z} = \frac{\int_V \tilde{z}\, dV}{\int_V dV}. \tag{9.2}$$

  - **Area**. $C\ (\bar{x}, \bar{y}, \bar{z})$ is given by

$$\bar{x} = \frac{\int_A \tilde{x} dA}{\int_A dA}, \quad \bar{y} = \frac{\int_A \tilde{y}\, dA}{\int_A dA}, \quad \bar{z} = \frac{\int_A \tilde{z}\, dA}{\int_A dA}. \tag{9.3}$$

  - **Line**. $C\ (\bar{x}, \bar{y}, \bar{z})$ is given by

$$\bar{x} = \frac{\int_L \tilde{x}\, dL}{\int_L dL}, \quad \bar{y} = \frac{\int_L \tilde{y}\, dL}{\int_L dL}, \quad \bar{z} = \frac{\int_L \tilde{z} dL}{\int_L dL}. \tag{9.4}$$

- The centroid will lie on any axis of symmetry of the body. Also, the centroid may be located off the body e.g., in the case of a ring where the centroid is at the center.

9.2 COMPOSITE BODIES

A composite body consists of a series of connected "simpler" shaped bodies.

PROCEDURE FOR ANALYSIS

The location of the center of gravity of a composite body can be determined using the following procedure:

- **Composite Parts**
  - ◆ Using a sketch, divide the body or object into a finite number of composite parts that have simpler shapes.
  - ◆ If a composite part has a *hole*, then consider the composite part without the hole and consider the hole as an *additional* composite part having *negative* weight or size.
- **Moment Arms.**
  - ◆ Establish the coordinate axes on the sketch and determine the coordinates $(\tilde{x}, \tilde{y}, \tilde{z})$ of the center of gravity of each composite part.

- **Summations**
  - Determine $\bar{x}$, $\bar{y}$, $\bar{z}$ by applying the center of gravity equations:

$$\bar{x} = \frac{\sum \tilde{x} W}{\sum W}, \quad \bar{y} = \frac{\sum \tilde{y} W}{\sum W}, \quad \bar{z} = \frac{\sum \tilde{z} W}{\sum W}, \tag{9.5}$$

    where $\sum W$ is the sum of the weights of all the composite parts of the body (total weight of the body).
  - If an object is *symmetrical* about an axis, the centroid of the object lies on this axis.

- **CENTROID FOR A COMPOSITE**—When the (composite) body has *constant* density or specific weight, the center of gravity *coincides* with the centroid of the body which, for lines, areas and volumes, can be found using relations analogous to (9.5) with the W's replaced by L's, A's and V's, respectively [as in (9.2) - (9.4)]. Centroids for common shapes of lines, areas, shells and volumes that often make up a composite body are given in the table on the inside back cover of the text.

9.3 THEOREMS OF PAPUS AND GULDINUS

The following two theorems (of Papus and Guldinus) are used to find the *surface area and volume* of any object of revolution:

- **Surface Area**. The area A of a surface of revolution equals the product of the length of the generating curve and the distance travelled by the centroid of the curve in generating the surface area. That is:

$$A = \theta \bar{r} L,$$

  where θ is the angle of revolution (radians), $\bar{r}$ is the perpendicular distance from the axis of revolution to the centroid of the generating curve and L is the total length of the generating curve.
- **Volume**. The volume V of a body of revolution equals the product of the generating area and the distance traveled by the centroid of the area in generating the volume. That is:

$$V = \theta \bar{r} A,$$

  where θ is the angle of revolution (radians), $\bar{r}$ is the perpendicular distance from the axis of revolution to the centroid of the generating area and A is the generating area.
- **Composite Shapes**. These two theorems may also be applied to lines or areas that may be composed of a series of composite parts. In this case, the total surface area or volume generated is the sum of the surface areas of volumes generated by each of the composite parts:

$$A = \theta \sum \tilde{r} L, \quad V = \theta \sum \tilde{r} A,$$

  where $\tilde{r}$ is the distance from the axis of revolution to the centroid of each composite part (remember that each part undergoes the same angle of revolution θ).

9.4 RESULTANT OF A GENERAL DISTRIBUTED LOADING

In Section 4.9 we discussed the method used to simplify a distributed loading which is uniform along an axis of a rectangular surface. Here, we generalize this method to include surfaces which have an arbitrary shape and are subjected to a variable load distribution.

– **Pressure Distribution over a Surface.** Consider a flat plate subjected to the loading function $p(x, y)$ Pa ($Pa = 1\text{N/m}^2$). The entire loading on the plate can be simplified to a *single resultant force* $\mathbf{F}_R$

- *Magnitude of Resultant Force.*

$$FR = \int V \, dV$$

i.e., total volume under the distributed loading diagram.

- *Location of Resultant Force.* The location $(\bar{x}, \bar{y})$ of $\mathbf{F}_R$ is given by

$$\bar{x} = \frac{\int_V x \, dV}{\int_V dV}, \quad \bar{y} = \frac{\int_V y \, dV}{\int_V dV}. \tag{9.6}$$

In other words,

Line of Action of $\mathbf{F}_R$ passes through the geometric center or centroid of the volume under the distributed loading diagram i.e.

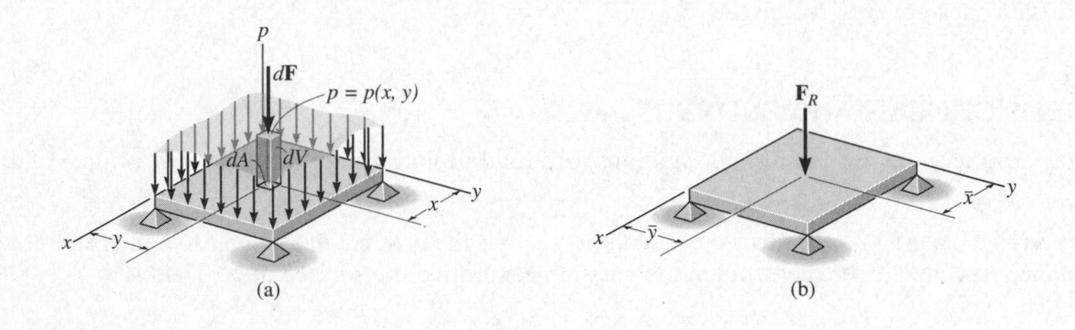

9.5 FLUID PRESSURE

According to *Pascal's law* a fluid at rest creates a pressure p at a point that is the *same in all* directions. The magnitude of p (force per unit area) is given by

$$p = \gamma z = \rho g z, \tag{9.7}$$

where γ is specific weight, ρ is mass density and z is the depth of the point from the fluid surface. Equation (9.7) is valid only for *incompressible fluids* (in which pressure and temperature variations do not produce any significant density variations).

Using Equation (9.7) and the results of Section 9.4, it is possible to determine the resultant force caused by a liquid pressure distribution and specify its location on the surface of a submerged plate. Three cases are considered:

- **Flat Plate of Constant Width.** The easiest of the three cases. Consider a flat rectangular plate of constant width submerged in a liquid with specific weight γ. The magnitude of the resultant force $\mathbf{F}_R$ resulting from the distribution of pressure over the plate's surface is equal to the trapezoidal volume having an intensity of $p_1 = \gamma z_1$ at depth z_1 and $p_2 = \gamma z_2$ at depth z_2. The line of action of $\mathbf{F}_R$ passes through the *volume's centroid C* [see Equation (9.6)]—this is *not* the centroid of the plate but rather the *center of pressure P of the plate.*
- **Curved Plate of Constant Width.** The calculation of the magnitude of $\mathbf{F}_R$ and its location P is more complicated for a (general) curved plate than a flat plate. There is a simplification, however, when the plate has *constant width.* This method requires separate calculations for the horizontal and vertical *components* of $\mathbf{F}_R$.
- **Flat Plate of Variable Width.** The loading caused by the pressure distribution acting on the surface of a submerged plate having a variable width has resultant $\mathbf{F}_R$ with *magnitude given by the volume described by the plate area as its base and linearly varying pressure distribution as its height.* From Equation (9.6), the centroid of V again defines the point through which $\mathbf{F}_R$ acts i.e., the center of pressure P, which lies on the surface of the plate just below C, has coordinates $P\left(\bar{x}, \bar{y}'\right)$ defined by the equations

$$\bar{x} = \frac{\int V \tilde{x} \, dV}{\int V \, dV}, \quad \bar{y}' = \frac{\int V \tilde{y}' \, dV}{\int V \, dV}.$$

Note that this point is *not* the centroid of the plate's area.

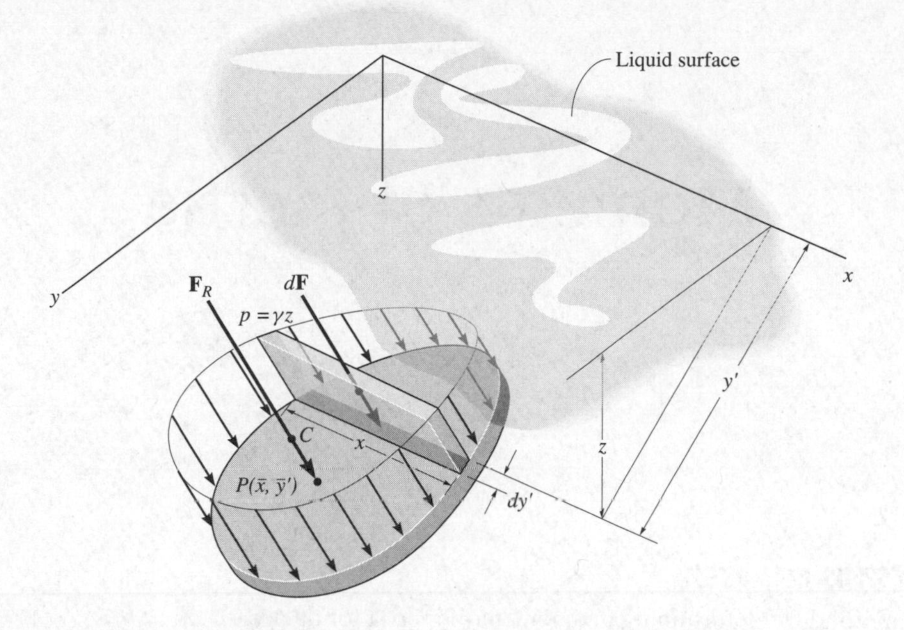

HELPFUL TIPS AND SUGGESTIONS

- Finding the location of the center of mass or centroid involves multiple integration (e.g., over volumes or surfaces). These integrations can often be reduced to single integrations using the procedure outlined at the end of Section 9.1. Study this procedure then read and perform Examples 9-1 through 9-8 of the text.

REVIEW QUESTIONS

1. True or False?
   (a) The location of the center of mass coincides with that of the center of gravity.
   (b) The centroid of a body always coincides with the body's center of mass.
   (c) The centroid of a body is always located on the body in question.
   (d) The formula for the centroid for the surface area of a plate involves integrals over the same surface area.
2. If the total weight of an object is to be represented by a single equivalent force, where must this force act?
3. What's the difference between the center of gravity and the center of mass?
4. How is the specific weight of a body defined?
5. What's the relationship between the mass density ρ and the specific weight γ of a body?
6. What does it mean when we say a body is homogeneous?
7. If an object is homogeneous, what do you know about the position of its center of mass?
8. In the analysis of the location of the center of gravity of a composite body, how do you deal with a hole in the body?
9. What are the theorems of Papus and Guldinus used for?
10. Show that the centroid for the volume of a body coincides with the *center of mass* only if the material composing the body is *uniform or homogeneous.*

10

Moments of Inertia

MAIN GOALS OF THIS CHAPTER:

- To develop a method for determining the moment of inertia for an area.
- To introduce the product of inertia and show how to determine the maximum and minimum moments of inertia of an area.
- To discuss the mass moment of inertia.

10.1 DEFINITION OF MOMENTS OF INERTIA FOR AREAS

- The moments of inertia (second moments) of the area A about the x and y axes are given, respectively, by

$$I_x = \int_A y^2 \, dA, \tag{10.0}$$

$$I_y = \int_A x^2 \, dA. \tag{10.1}$$

- The moment of inertia of the area A about the pole O or the z-axis (also known as the *polar moment of inertia*) is

$$J_O = \int_A r^2 \, dA = I_x + I_y \tag{10.2}$$

where $r^2 = x^2 + y^2$ is the perpendicular distance from the pole (z-axis) to the element dA.

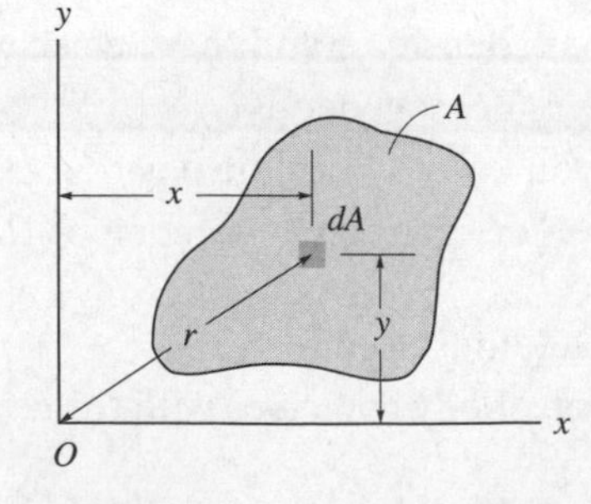

- Clearly I_x, I_y and J_O are always positive and have units of length raised to the fourth power.
- The terminology "moment of inertia" is actually a misnomer in this context—it has been adopted because of the similarity with integrals of the same form related to *mass*.

10.2 PARALLEL AXIS THEOREM FOR AN AREA

- Suppose the centroid of an area is located at $C\left(x', y', z'\right)$. *The moment of inertia of an area A about an axis is equal to the moment of inertia of the area about a parallel axis* **passing through the area's centroid** *plus the product of the area and the square of the perpendicular distance between the axes.* e.g.

$$I_x = \bar{I}_{x'} + Ad_y^2, \quad I_y = \bar{I}_{y'} + Ad_x^2, \quad J_O = \bar{J}_C + Ad^2 \tag{10.3}$$

where $\bar{I}_{x'}$, $\bar{I}_{y'}$ and $\bar{J}_C$ represent moments of inertia of the area about a corresponding parallel axis passing through the area's centroid.

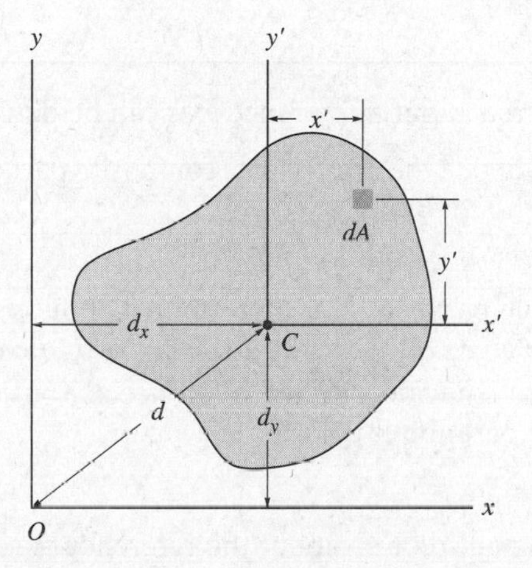

10.3 RADIUS OF GYRATION OF AN AREA

- Provided the areas and moments of inertia are known, the *radii of gyration* of a planar area are determined from the formulas

$$k_x = \sqrt{\frac{I_x}{A}}, \quad k_y = \sqrt{\frac{I_y}{A}}, \quad k_O = \sqrt{\frac{J_O}{A}} \quad \text{(units of length).} \tag{10.4}$$

- It is clear from equations (10.0)–(10.1) that finding moments of inertia requires the evaluation of *area integrals*. If one chooses to describe the area element dA with differential size in two directions (e.g., $dA = dx\,dy$) then a *double integration* must be performed. Most often, however, it is easier to perform only a single integration by *specifying the differential element dA* having a differential size or thickness in only one direction. The procedure is as follows:

PROCEDURE FOR ANALYSIS

- Most often the element dA can be rectangular with a *finite length* and differential width.
- The element should be located so that it intersects the boundary of the area at the arbitrary point (x, y). There are two ways to orient the element dA with respect to the axis about which the moment of inertia is to be determined:
  - ◆ **Case 1**: The *length* of the element can be oriented *parallel* to the axis. In this case, Equations (10.0)–(10.1) can be used *directly* since *all parts* of the element lie at the *same* moment-arm distance from the axis.
  - ◆ **Case 2**: The *length* of the element can be oriented *perpendicular* to the axis. Here neither of Equations (10.0) - (10.1) can be used directly since all parts of the element will not lie at the same moment-arm distance from the axis. Instead, it is necessary to first calculate the moment of inertia of the *element* (separately) and then integrate this result over the area A to obtain the appropriate area moment of inertia.
- See Examples 10-1 to 10-4 in text.

10.4 MOMENTS OF INERTIA FOR COMPOSITE AREAS

Provided the moment of inertia of each of the simpler areas (making up the composite area) is known, or can be determined about a common axis, then the moment of inertia of the composite area equals the *algebraic sum* of the moments of inertia of all its parts.

PROCEDURE FOR ANALYSIS

The moment of inertia of a composite area about a reference axis can be determined using the following procedure:

- **Composite Parts**—Using a sketch, divide the area into its composite parts and indicate the perpendicular distance from the centroid of each part to the reference axis.
- **Parallel-Axis Theorem.**
  - The moment of inertia of each part should be determined about its centroidal axis, which is parallel to the reference axis—*use the table given on the inside back cover of the text.*
  - If the centroidal axis does not coincide with the reference axis, use the parallel axis theorem to determine the moment of inertia of the part about the reference axis.
- **Summation.**
  - The moment of inertia of the entire area about the reference axis is found by summing the results of the composite parts.
  - If a composite part has a "hole," its moment of inertia is found by "subtracting" the moment of inertia for the hole from the moment of inertia of the entire part including the hole.

10.5 PRODUCT OF INERTIA FOR AN AREA

In general, the moment of inertia for an area is different for every axis about which it is computed. In some applications, it is necessary to know the orientation of those axes which give, respectively, the maximum and minimum moments of inertia for the area (see Section 10.6). Essential to this is the idea of a *product of inertia for an area.*

- The product of inertia for the area A is

$$I_{xy} = \int_A xy\,dA$$

  - The product of inertia may be negative, positive or zero (unlike moment of inertia). For example I_{xy} will be zero if x or y is an axis of symmetry for the area A.
  - The sign of I_{xy} depends on the quadrant where the area A is located. In fact, if the area is rotated from one quadrant to another, the sign of I_{xy} will change.

- **Parallel Axis Theorem for Product of Inertia of an Area** A

$$I_{xy} = \bar{I}_{x'y'} + A d_x d_y$$

It is important that the *algebraic signs* for d_x and d_y be maintained when applying this result.

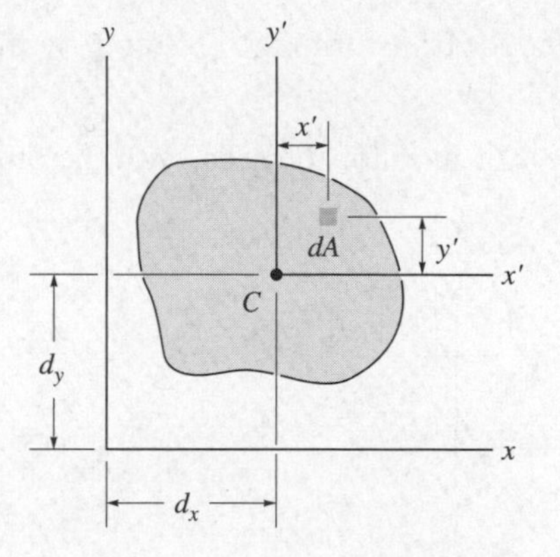

10.6 MOMENTS OF INERTIA FOR AN AREA ABOUT INCLINED AXES

- The moments and products of inertia for an area with respect to a set of inclined (at an angle θ) u and v axes are given by (assuming θ, I_x, I_y and I_{xy} are *known*):

$$I_u = \frac{I_x + I_y}{2} + \frac{I_x - I_y}{2}\cos 2\theta - I_{xy}\sin 2\theta,$$
$$I_v = \frac{I_x + I_y}{2} - \frac{I_x - I_y}{2}\cos 2\theta + I_{xy}\sin 2\theta, \tag{10.5}$$
$$I_{uv} = \frac{I_x - I_y}{2}\sin 2\theta + I_{xy}\cos 2\theta$$

- The polar moment of inertia about the z-axis passing through the point O is independent of the orientation of the u and v axes i.e.

$$J_O = I_u + I_v = I_x + I_y$$

PRINCIPAL MOMENTS OF INERTIA

- When the angle θ in (10.5) takes the value $\theta = \theta_p$ defined by

$$\tan 2\theta_p = \frac{-I_{xy}}{(I_x - I_y)/2}, \tag{10.6}$$

the axes u and v are called the *principal axes of the area* since they identify the orientation of the axes u and v about which the moments of inertia I_u and I_v are *maximum or minimum*. In this case, they are called *principal moments of inertia* and are given by

$$I_{\substack{\max\\ \min}} = \frac{I_x + I_y}{2} \pm \sqrt{\left(\frac{I_x - I_y}{2}\right)^2 + I_{xy}^2}. \tag{10.7}$$

Depending on the sign chosen, this result gives the maximum or minimum moment of inertia for the area.

- The *product of inertia with respect to the principal axes is zero*. Hence any symmetrical axis represents a principal axis of inertia for the area.

10.7 MOHR'S CIRCLE FOR MOMENTS OF INERTIA

Equations 10.5 to 10.7 have a graphical solution which is convenient to use and easy to remember—this solution is called a *Mohr's circle.*

PROCEDURE FOR ANALYSIS

Mohr's circle provides a convenient means for transforming I_x, I_y and I_{xy} into the principal moments of inertia using the following procedure:

- **Determine** I_x, I_y, I_{xy}. Establish the x, y axes for the area, with the origin located at the point P of interest and determine I_x, I_y and I_{xy}.

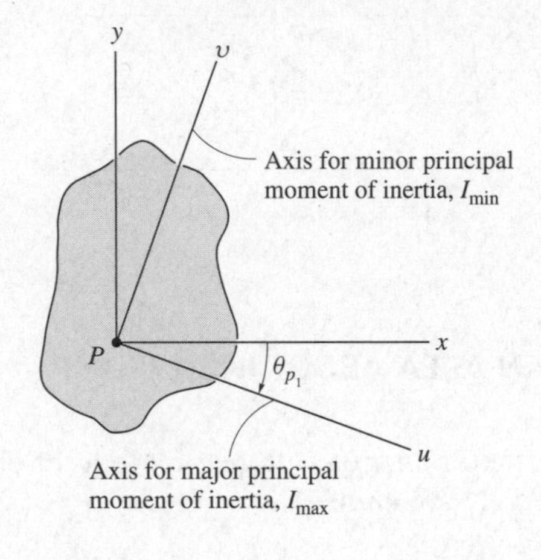

- **Construct the Circle**
  - Construct a rectangular coordinate system such that the abscissa represents the moment of inertia I and the ordinate represents the product of inertia I_{xy}
  - Determine the center of the circle O, which is located at a distance $\dfrac{I_x + I_y}{2}$ from the origin and plot the reference point A having coordinates (I_x, I_{xy}). By definition, I_x is always positive, whereas I_{xy} will be either positive or negative.
  - Connect the reference point A with the center of the circle and determine the distance OA by trigonometry. This distance represents the radius of the circle. Finally draw the circle.

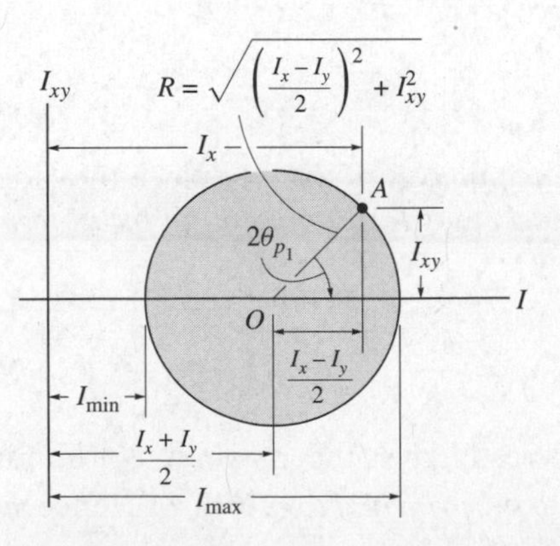

- **Principal Moments of Inertia.** The points where the circle intersects the abscissa give the values of the principal moments of inertia $I_{\min}$ and $I_{\max}$. Notice that the *product of inertia will be zero at these points.*
- **Principal Axes.** To find the direction of the major principal axis, determine, by trigonometry, the angle $2\theta_{p_1}$, measured from the radius OA to the positive I-axis. This angle represents twice the angle from the x axis of the area in question to the axis of maximum moment of inertia $I_{\max}$. Both the angle on the circle, $2\theta_{p_1}$ and the angle to the axis on the area, θ_{p_1}, must be measured in the same sense. The axis for minimum moment of inertia $I_{\min}$ is perpendicular to the axis for $I_{\max}$.

10.8 MASS MOMENT OF INERTIA

- The mass moment of inertia about the z-axis is given by

$$I = \int_m r^2 \, dm \tag{10.8}$$

where r is the perpendicular distance from the axis to the arbitrary element dm.

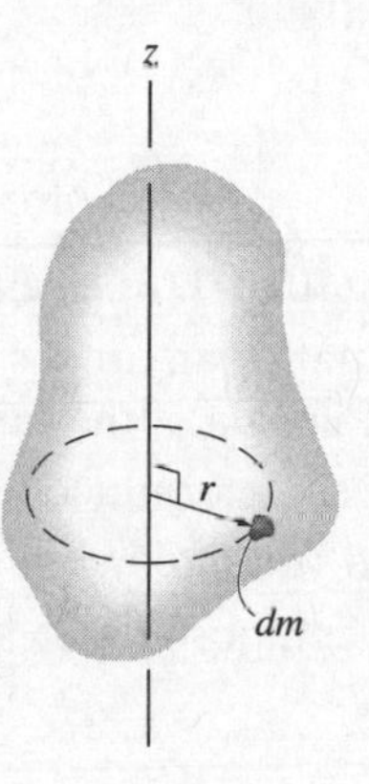

- When the axis passes through the body's mass center G, the moment of inertia is denoted by I_G. The mass moment of inertia is always positive and has units $kg.m^2$ or $slug.ft^2$.
- If the body consists of a material having a variable density $\rho\,(x, y, z)$, the body's moment of inertia is computed using volume integration as

$$I = \int_V r^2 \rho \, dV.$$

This integral is generally computed as a *triple integral.* The integration process can, however, be simplified to a *single integration* provided the chosen elemental volume has a differential size or thickness in only *one* direction. Shell or disk elements are often used for this purpose.

PARALLEL AXIS THEOREM

- *If the moment of inertia of the body about an axis passing through the body's mass center is known, then the moment of inertia about any other parallel axis can be computed from*

$$I = I_G + md^2 \tag{10.9}$$

where m is the mass of the body and d is the perpendicular distance between the axes.

RADIUS OF GYRATION

- The radius of gyration k has units of length and is related to the mass m and moment of inertia I of the body by

$$I = mk^2 \quad \text{or} \quad k = \sqrt{\frac{I}{m}}$$

COMPOSITE BODIES

- If a body is constructed from a number of simple shapes such as disks, spheres and rods, the moment of inertia of the body about any axis z can be determined by adding algebraically the moments of inertia of all the composite shapes computed about the z-axis. A composite part must be considered as a negative quantity if it has already been included within another part e.g., a "hole" subtracted from a solid plate.

HELPFUL TIPS AND SUGGESTIONS

- Be careful when using the Parallel Axis Theorem (10.9). It is applicable only to the situation when the moment of inertia of the body about an axis *passing through the body's mass center* is known. It *cannot* be applied in the form $I = I_B + md^2$ where B is an arbitrary point.

REVIEW QUESTIONS

1. True or False? The area moments of inertia (10.0)–(10.2) can be negative.
2. It is often the case that the moment of inertia for an area is known about an axis passing through its centroid. What can then be said about the moment of inertia of the same area about a corresponding parallel axis?
3. How are the radii of gyration for a planar area determined?
4. If a composite body has a "hole," how would you find moment of inertia of the body ?
5. Find the moment of inertia of the following composite area about the x-axis.

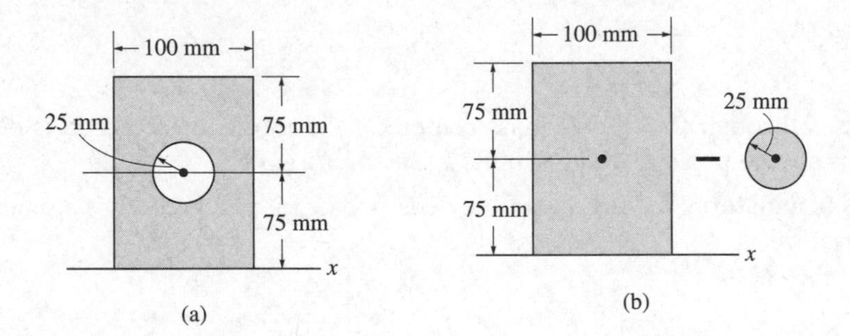

6. What are the *principal axes of an area* and the *principal moments of inertia*?
7. What are *Mohr's circles* used for?
8. Define the mass moment of inertia and describe how it can be found.

11

Virtual Work

MAIN GOALS OF THIS CHAPTER:

- To introduce the principle of *virtual work* and show how it applies to determining the equilibrium configuration of a series of pin-connected members.
- To establish the potential energy function and use the potential-energy method to investigate the type of equilibrium or stability of a rigid body or configuration.

11.1 DEFINITION OF WORK AND VIRTUAL WORK

WORK OF A FORCE

- A force **F** does work only when it undergoes a displacement in the direction of the force.
- Work is a *scalar quantity* defined by the dot product

$$dU = \mathbf{F} \cdot \mathbf{dr} = F \cos\theta \, ds, \tag{11.0}$$

  where dU is the increment of work done when the force **F** is displaced **dr**, θ is the angle between the tails of **dr** and **F**, and ds is the magnitude of **dr**.
- *Positive work* is done when the force and its displacement have the same *sense*. Otherwise *negative work* is done.
- In the SI system, the basic unit of work is a *Joule* (J) ($1\ \text{J} = 1\ \text{N} \cdot \text{m}$). In the FPS system work is defined in terms of ft.lb.

WORK OF A COUPLE

- The two forces of a couple do work when the couple *rotates* about an axis perpendicular to the plane of the couple.
- Work done by a couple **M** is a *scalar quantity* defined by the dot product

$$dU = \mathbf{M} \cdot d\boldsymbol{\theta} \tag{11.1}$$

where dU is the increment of work done when the couple **M** is 'displaced' $\mathbf{d\theta}$ (a differential rotation of the body about an axis perpendicular to the plane of the couple).

- The resultant work is *positive* when the sense of **M** is the same as that of $\mathbf{d\theta}$ and *negative* when they have an *opposite sense*. The direction and sense of $\mathbf{d\theta}$ are again defined by the *right-hand rule*. Hence, if movement of the body occurs in the *same plane*, the line of action of $\mathbf{d\theta}$ will be *parallel* to the line of action of **M** and Equation (11.1) becomes

$$dU = Md\theta \tag{11.2}$$

VIRTUAL WORK

- A *virtual* movement (displacement or rotation) is an *imaginary* movement which is *assumed* and *does not actually exist*. A *virtual displacement* is a differential that is given in the positive direction of the position coordinate and is denoted by the symbol δs. Similarly, a *virtual rotation* is denoted by $\delta\theta$.
- The *virtual work* done by a force undergoing a virtual displacement δs is

$$\delta U = F\cos\theta\delta s. \tag{11.3}$$

- The *virtual work* done by a couple undergoing a virtual rotation $\delta\theta$ in the plane of the couple forces is

$$\delta U = M\delta\theta. \tag{11.4}$$

11.2 PRINCIPLE OF VIRTUAL WORK FOR A PARTICLE AND A RIGID BODY

- **Particle.** If a particle undergoes an imaginary or virtual displacement $\delta\mathbf{r}$, then the virtual work (δU) done by the force system must be zero for equilibrium i.e.

$$\delta U = 0. \tag{11.5}$$

In other words we can write *three independent virtual work equations* corresponding to the three equations of equilibrium:

$$\sum Fx\delta x = 0, \quad \sum Fy\delta y = 0, \quad \sum Fz\delta z = 0.$$

- **Rigid Body**. As in the case of a particle we can also write a set of three virtual work equations (11.5) for a rigid body subjected to a coplanar force system (two involving virtual translations in the x and y directions and another a virtual rotation about an axis perpendicular to the x–y plane and passing through an arbitrary point O).
- **NOTE.** As in the case of a particle, no added advantage is gained by solving rigid body equilibrium problems using the principle of virtual work since for each application of the virtual work equation, the virtual displacement or rotation, common to every term, factors out, leaving an equation that could have been obtained in a more direct manner by simply applying the equations of equilibrium.

11.3 PRINCIPLE OF VIRTUAL WORK FOR A SYSTEM OF CONNECTED RIGID BODIES

The method of virtual work is most suitable for solving equilibrium problems that involve a system of several *connected* rigid bodies.

- **Degrees of Freedom.** An *n–degree-of-freedom system* requires *n independent coordinates* q_n to specify the location of all its members.

- **Principle of Virtual Work.** The principle of virtual work for a system of rigid bodies whose connections are *frictionless* may be stated as follows:

  A system of connected rigid bodies is in equilibrium provided the virtual work done by all the external forces and couples acting on the system is zero for each independent virtual displacement of the system

  Mathematically, we write

$$\delta U = 0 \tag{11.6}$$

  where δU represents the virtual work of all the external forces and couples acting on the system during any independent virtual displacement.
- This means that for an $n-$ degree-of-freedom system it is possible to write n independent virtual work equations, one for every virtual displacement taken along each of the independent coordinate axes, while the remaining $n - 1$ remaining independent coordinates are held *fixed.*

The following procedure shows how to use the equation of virtual work to solve problems involving a system of frictionless connected rigid bodies having a *single degree of freedom.*

- **Free-Body Diagram**
  - Draw the free-body diagram of the entire system of connected bodies and define the independent coordinate q.
  - Sketch the "deflected position" of the system on the free-body diagram when the system undergoes a positive virtual displacement δq.
- **Virtual Displacements**
  - Indicate position coordinates s_i, measured from a *fixed point* on the free-body diagram to each of the i number of "active" forces and couples i.e., those that do work.
  - Each coordinate axis should be parallel to the line of action of the "active' force to which it is directed, so that the virtual work along the coordinate axis can be calculated.
  - Relate each of the position coordinates s_i to the independent coordinate q; then differentiate these expressions to express the virtual displacements δs_i in terms of δq.
- **Virtual Work Equation**
  - Write the virtual work equation (11.6) for the system assuming that, whether possible or not, all the position coordinates s_i undergo *positive* virtual displacements δs_i.
  - Using the relations for δs_i, express the work of *each*"active" force and couple in the equation in terms of the single independent virtual displacement δq.
  - Factor out the common displacement from all the terms and solve for the unknown force, couple or equilibrium position, q.
  - If the system contains n degrees of freedom, n independent coordinates q_n must be specified. Follow the above procedure and let *only one* of the independent coordinates undergo a virtual displacement while the remaining $n - 1$ coordinates are *held fixed.* In this way, n virtual-work equations can be written, one for each independent coordinate.

- **EXAMPLES.** The above procedure is illustrated in Examples 11-1 through 11-4 in the text. *Note that had these examples been solved using the equations of equilibrium, it would have been necessary to dismember the links and apply three scalar equations to each link. The principle of virtual work, by means of calculus, has eliminated this task so that the answer is obtained directly.*

11.4 CONSERVATIVE FORCES

- If a force **F** is displaced over a path with finite length S the work done by the force is given by the integral

$$U = \int_S dU = \int_S F \cos\theta ds.$$

  If this integral is *independent of its path* (depends only on the initial and final locations of its path), the force is called a *conservative force.*
- **EXAMPLES OF CONSERVATIVE FORCES:** *Weight, Elastic Springs.*
- **EXAMPLES OF NONCONSERVATIVE FORCES:** *Friction:* the work done by the frictional force *depends on the path*: the longer the path, the greater the work. The work done is dissipated from the body in the form of heat.

11.5 POTENTIAL ENERGY

When a conservative force acts on a body, it gives the body the capacity to do work. This capacity is known as the body's *potential energy* and depends on the location of the body.

- **Gravitational Potential Energy.** Measuring *y positive upward*, the gravitational potential energy of a body's weight **W** is

$$V_g = Wy. \tag{11.7}$$

- **Elastic Potential Energy.** The elastic potential energy V_e that a spring produces on an attached body, when the spring is elongated or compressed from an undeformed position ($s = 0$) to a final position s is

$$V_e = \frac{1}{2}ks^2. \tag{11.8}$$

- **Potential Function.** In the general case, if a body is subjected to both gravitational and elastic forces, the potential energy (function) V of the body can be expressed as the algebraic sum

$$V = V_g + V_e$$

  where measurement of V depends on the location of the body with respect to a selected datum in accordance with Equations (11.7) and (11.8).

11.6 POTENTIAL ENERGY CRITERION FOR EQUILIBRIUM

- **System Having One Degree of Freedom (q).** When a frictionless connected system of rigid bodies is in equilibrium, we require that the potential energy (function) V of the body satisfies

$$\frac{dV}{dq} = 0. \tag{11.9}$$

- **System Having n Degrees of Freedom ($q_1, \ldots, q_n$).** When a frictionless connected system of rigid bodies is in equilibrium, we require that the potential energy (function) V of the body satisfies

$$\frac{\partial V}{\partial q_1} = 0, \quad \frac{\partial V}{\partial q_2} = 0, \ldots, \frac{\partial V}{\partial q_n} = 0.$$

  In other words, it is possible to *write n independent equations for a system having n degrees of freedom.*

11.7 STABILITY OF EQUILIBRIUM

Once the equilibrium configuration for a body or a system of connected bodies is defined, it is important to investigate the "type" of equilibrium or the stability of the configuration.

- **Types of Equilibrium.**
  1. *Stable Equilibrium.* A small displacement of the system causes the system to return to its original position. Potential energy of the system is at a minimum in this case.
  2. *Neutral Equilibrium.* A small displacement of the system causes the system to remain in its displaced state. Potential energy of the system remains constant in this case.
  3. *Unstable Equilibrium.* A small displacement of the system causes the system to move farther away from its original position. Original potential energy of the system is a maximum in this case.

- **System Having One Degree of Freedom (q).** We require that the potential energy (function) V of the body satisfies the following conditions in each case:
  1. *Stable Equilibrium.*

$$\frac{dV}{dq} = 0, \quad \frac{d^2V}{dq^2} > 0. \tag{11.10}$$

  2. *Neutral Equilibrium.*

$$\frac{dV}{dq} = \frac{d^2V}{dq^2} = \frac{d^3V}{dq^3} = \cdots = 0. \tag{11.11}$$

  3. *Unstable Equilibrium.*

$$\frac{dV}{dq} = 0, \quad \frac{d^2V}{dq^2} < 0. \tag{11.12}$$

- **System Having Two Degrees of Freedom (q_1, q_2).** Things become much more complicated as the number of degrees of freedom of the system increases. However, for a system for two degrees of freedom, we can say:
  1. *Equilibrium and Stability* occur at a point (q_{1eq}, q_{2eq}) when

$$\frac{\partial V}{\partial q_1} = \frac{\partial V}{\partial q_2} = 0,$$

$$[\left(\frac{\partial^2 V}{\partial q_1 \partial q_2}\right)^2 - \left(\frac{\partial^2 V}{\partial q_1^2}\right)\left(\frac{\partial^2 V}{\partial q_2^2}\right)] < 0,$$

$$(\frac{\partial^2 V}{\partial q_1^2} + \frac{\partial^2 V}{\partial q_2^2}) > 0.$$

  2. *Equilibrium and Instability* occur when

$$\frac{\partial V}{\partial q_1} = \frac{\partial V}{\partial q_2} = 0,$$

$$[\left(\frac{\partial^2 V}{\partial q_1 \partial q_2}\right)^2 - \left(\frac{\partial^2 V}{\partial q_1^2}\right)\left(\frac{\partial^2 V}{\partial q_2^2}\right)] < 0,$$

$$\left(\frac{\partial^2 V}{\partial q_1^2} + \frac{\partial^2 V}{\partial q_2^2}\right) < 0.$$

PROCEDURE FOR SOLVING PROBLEMS

Using potential-energy methods, the equilibrium positions and the stability of a body or a system of connected bodies having a *single degree of freedom* q can be obtained using the following procedure.

- **Potential Function.**
  - Sketch the system so that it is located at some *arbitrary position* specified by the independent coordinate q.
  - Establish a horizontal *datum* through a *fixed point* and express the *gravitational potential energy* V_g in terms of the weight W of each member and its vertical distance y from the datum, $V_g = Wy$.
  - Express the elastic potential energy V_e of the system in terms of the stretch or compression, s, of any connecting spring and the spring stiffness k, $V_e = \frac{1}{2}ks^2$.
  - Formulate the potential function $V = V_g + V_e$ and express the *position coordinates* y and s in terms of the independent coordinate q.
- **Equilibrium Position.**
  - The equilibrium position is determined from Equation (11.9) i.e. $\frac{dV}{dq} = 0$.
- **Stability**. Stability at the equilibrium position is determined from Equations (11.10)–(11.12).

REVIEW QUESTIONS

1. What is the work done by a force $\mathbf{F}$ when its point of application is displaced $\mathbf{dr}$?
2. What is the work done by a couple $\mathbf{M}$ when the object on which it acts rotates through an angle $\mathbf{d\theta}$?
3. What does the principle of virtual work say when an object in equilibrium is subjected to a virtual translation or rotation?
4. What is meant by a "conservative force"?
5. What is the potential energy of a body and how is it related to the concept of "conservative force"?
6. What is the potential energy criterion for equilibrium for a frictionless connected system of rigid bodies with one degree of freedom?
7. What does it mean when an equilibrium position of a body is stable or unstable?
8. How do you know when an equilibrium position of a system having one degree of freedom is stable or unstable?

ANSWERS TO REVIEW QUESTIONS

Chapter 1:

1. F **2.** T **3.** F **4.** F **5.** F **6.** T **7.** T **8.** F

Chapter 2:

1. See (2.1) **2.** See (2.0) (in the plane) or (2.2) and (2.3) **3.** See (2.4)

4. See Section 2.5 **5.** See (2.3) **6.** See (2.5)

7. Vectors are perpendicular.

8. See (2.5) **9.** See (2.6) **10.** See (2.7) and (2.8)

Chapter 3:

1. See Section 3.1 **2.** See (3.0) **3.** See Section 3.2

4. Lines of action of the forces lie in a plane

5. Lines of action of the forces lie in three-dimensional space

6. One more equation—see (3.2) and (3.3)

7. See (3.1) **8.** True.

Chapter 4:

1. See Section 4.1. **2.** See Section 4.1. **3.** Right-hand rule.

4. No moment. **5.** See (4.0). **6.** Vectors are parallel.

7. **r** represents a position vector drawn *from O to any point* lying on the line of action of **F**.

8. $|\mathbf{M}_O| = |\mathbf{F}|\,|\mathbf{r}| \sin\theta = Fr\sin\theta = Fd$.

9. Right-hand rule i.e. curling the fingers of the right hand from vector **r** (cross) to vector **F**, the thumb then points in the direction of $\mathbf{M}_O$.

10. See Section 4.6. **11.** False. **12.** See Section 4.7. **13.** See Section 4.7 and (4.2).

14. See Section 4.8. **15.** See Section 4.9.

Chapter 5:

1. A particle has no size/shape and so cannot support rotation, only translation.

2. See (5.1) and (5.0).

3. See Section 5.7: the object has more supports than are necessary to hold it in equilibrium

4. See Section 5.7: when there are more unknown loadings on the body than equations of equilibrium available for their solution.

5. Review Section 5.7.

Chapter 6:

1. See Section 6.1: A *truss* is a structure composed of slender members joined together at their end points.
2. See Section 6.1.
3. See Section 6.2: In order to analyze or design a truss, we must obtain the force in each of its members. To do this, we consider the *equilibrium of a joint* of the truss. This is the basis of the method of joints.
4. Two
5. See Section 6.4.
6. See Section 6.5—either the *method of joints* or *method of sections*.
7. See Section 6.6.
8. The forces at the connected parts of the group are *internal* forces and are not shown on the free-body diagram *of the group*.

Chapter 7:

1. See Section 7.1: Normal force **N** acts parallel to the beam's axis. Shear force **V** acts normal to the beam's axis. Bending moment **M** is a couple moment which causes the beam to bend..
2. See Section 7.2. Follow the sign convention shown in the figure.
3. By integration to obtain V and M as functions of x.
4. See Section 7.4. Note that no cable is truly 'weightless,' 'inextensible' or 'perfectly flexible.' These terms are simplifications to aid the modeling.
5. We use equation (7.3) and integrate (noting that w is constant). We obtain

$$y\,(x) = \frac{1}{F_H}\left(\frac{wx^2}{2} + c_1 x + c_2\right)$$

This is a parabola. If the origin of the x–y coordinate system is chosen so that $y = 0$, $\dfrac{dy}{dx} = 0$ at $x = 0$, we obtain

$$y(x) = \frac{wx^2}{2F_H}$$

Chapter 8:

1. See Section 8.1: $F_s = \mu_s N$.
2. See Section 8.1: $F_k = \mu_k N$.
3. See Section 8.2: The total number of unknowns is equal to the total number of available equilibrium equations. In this case, once the frictional forces are determined, check that $F \leq \mu_s N$ otherwise slipping will occur and the body will not remain in equilibrium.
4. See Section 8.2: The total number of unknowns will *equal* the total number of available equilibrium equations plus the total number of available frictional equations or conditional equations for tipping.
5. Draw a free-body diagram!
6. Equation (8.0).
7. Equation (8.1) or (8.2) with $\phi = \phi_k$.
8. Equation (8.3) with R replaced by P (since the reactive force R is equal in magnitude to the load **P**).

Chapter 9:

1. (i) T (ii) F (material comprising the body must be homogeneous). (iii) F (iv) T—see (9.3).
2. See Section 9.1—at the center of gravity (or center of mass).
3. The *location* of the center of gravity *coincides* with that of the center of mass. However, the *center of mass is independent of gravity* and so can be used in situations when particles are not under the influence of a gravitational attraction—see Section 9.1.

4. Specific weight is γ or weight per unit volume.
5. $\gamma = \rho g$—see Section 9.1.
6. A body whose mass density is constant throughout its volume is said to be homogeneous—see Section 9.1.
7. Center of mass coincides with centroid and center of gravity—see Section 9.1.
8. See Section 9.2. If a composite part has a *hole*, then consider the composite part without the hole and consider the hole as an *additional* composite part having *negative* weight or size.
9. To find the *surface area and volume* of any object of revolution—see Section 9.3.
10. From (9.1) with $dW = \rho g\, dV$, the formula for the center of mass of a body is given by

$$\bar{x} = \frac{\int_V \tilde{x}\rho g dV}{\int_V \rho g dV}, \quad \bar{y} = \frac{\int_V \tilde{y}\rho g dV}{\int_V \rho g dV}, \quad \bar{z} = \frac{\int_V \tilde{z}\rho g dV}{\int_V \rho g dV}.$$

If the body is homogeneous ρ is constant. Thus we obtain:

$$\bar{x} = \frac{\int_V \tilde{x} dV}{\int_V dV}, \quad \bar{y} = \frac{\int_V \tilde{y} dV}{\int_V dV}, \quad \bar{z} = \frac{\int_V \tilde{z} dV}{\int_V dV}$$

which is exactly (9.2).

Chapter 10:

1. See Section 10.1: False.
2. See Section 10.2: *It's equal to the moment of inertia about the axis* **passing through the area's centroid** *plus the product of the area and the square of the perpendicular distance between the axes.*.
3. See Section 10.3, Equation (10.4).
4. See Section 10.4: If a composite part has a "hole," its moment of inertia is found by "subtracting" the moment of inertia for the hole from the moment of inertia of the entire part including the hole.
5. The composite area is determined by subtracting the circle from the rectangle. The centroid of each area is located in the figure.

- **Moment of inertia for Circle:** Using Equation (10.3)

$$\begin{aligned} I_x &= I_{x'} + Ad_y^2 \\ &= \frac{1}{4}\pi\,(25)^4 + \pi\,(25)^2\,(75)^2 = 11.4\,(10^6)\,\text{mm}^4 \end{aligned}$$

- **Moment of inertia for Rectangle:** Using Equation (10.3)

$$\begin{aligned} I_x &= I_{x'} + Ad_y^2 \\ &= \frac{1}{12}100\,(150)^3 + 100\,(150)^2\,(75)^2 = 112.5\,(10^6)\,\text{mm}^4 \end{aligned}$$

- **Summation.** The moment of inertia for the composite area is thus

$$\begin{aligned} I_x &= -11.4\,(10^6) + 112.5\,(10^6) \\ &= 101\,(10^6)\,\text{mm}^4 \end{aligned}$$

6. See Section 10.6: The *principal axes of the area* identify the orientation of the axes u and v about which the moments of inertia I_u and I_v are *maximum or minimum.* (the *principal moments of inertia)*.
7. See Section 10.7: Mohr's circle provides a convenient (graphical) means for transforming I_x, I_y and I_{xy} into the principal moments of inertia.
8. See Equation (10.8). This integral is generally computed as a volume and hence *triple integral (*using the relation $dm = \rho dV$). The integration process can, however, be simplified to a *single integration* provided the chosen elemental volume has a differential size or thickness in only *one* direction. Shell or disk elements are often used for this purpose. The detailed procedure is given in Section 10.8 of the text.

Chapter 11:

1. See (11.0). **2.** See (11.1).

4. That the virtual work (δU) done by the force system must be zero—see (11.5).

5. See Section 11.4—that the work done by the force over a finite path is independent of the path itself.

6. See Section 11.5. When a conservative force acts on a body, it gives the body the capacity to do work. This capacity is known as the body's *potential energy* and depends on the location of the body.

7. Equation (11.9).

8. See Section 11.7: *Stable Equilibrium:* a small displacement of the system causes the system to return to its original position. Potential energy of the system is at a minimum in this case; *Unstable Equilibrium.* A small displacement of the system causes the system to move farther away from its original position. Original potential energy of the system is a maximum in this case.

9. Test using Equation (11.10) for stability and Equation (11.12) for instability.

PART II

Free-Body Diagram Workbook

1

Basic Concepts in Statics

Statics is a branch of mechanics that deals with the study of bodies that are at rest (if originally at rest) or move with constant velocity (if originally in motion) that is, bodies which are in (static) equilibrium.

In mechanics, real bodies (e.g. planets, cars, planes, tables, crates, etc) are represented or *modeled* using certain idealizations which simplify application of the relevant theory. In this book we refer to only two such models:

- **Particle**. A *particle* has a mass but a size/shape that can be neglected. For example, the size of an aircraft is insignificant when compared to the size of the earth and therefore the aircraft can be modeled as a particle when studying its three-dimensional motion in space.
- **Rigid Body**. A *rigid body* represents the next level of sophistication after the particle. That is, a rigid body is a collection of particles which has a size/shape but this size/shape cannot change. In other words, when a body is modeled as a rigid body, we assume that any deformations (changes in shape) are relatively small and can be neglected. For example, the actual deformations occurring in most structures and machines are relatively small so that the rigid body assumption is suitable in these cases.

1.1 Equilibrium

Equilibrium of a Particle

A particle is in equilibrium provided it is at rest if originally at rest or has a constant velocity if originally in motion. To maintain equilibrium, it is necessary and sufficient to satisfy Newton's first law of motion which requires the resultant force acting on the particle or rigid body to be zero. In other words

$$\sum \mathbf{F} = \mathbf{0} \tag{1.1}$$

where $\sum \mathbf{F}$ is the vector sum of all the external forces acting on the particle.

Successful application of the equations of equilibrium (1.1) requires a complete specification of all the known and unknown external forces ($\sum \mathbf{F}$) that act on the object. The best way to account for these is to draw the object's *free-body diagram*.

Equilibrium of a Rigid Body

A rigid body will be in equilibrium provided the sum of all the external forces acting on the body is equal to zero and the sum of the external moments taken about a point is equal to zero. In other words

$$\sum \mathbf{F} = \mathbf{0} \tag{1.2}$$

$$\sum \mathbf{M}_O = \mathbf{0} \tag{1.3}$$

where $\sum \mathbf{F}$ is the vector sum of all the external forces acting on the rigid body and $\sum \mathbf{M}_O$ is the sum of the external moments about an arbitrary point O.

Successful application of the equations of equilibrium (1.2) and (1.3) requires a complete specification of all the known and unknown external forces ($\sum \mathbf{F}$) and moments ($\sum \mathbf{M}_O$) that act on the object. The best way to account for these is again to draw the object's *free-body diagram.*

2

Free-Body Diagrams: the Basics

2.1 Free-Body Diagram: Particle

The equilibrium equation (1.1) is used to determine unknown forces acting on an object (modeled as a particle) in equilibrium. The first step in doing this is to draw the *free-body diagram* of the object to identify the external forces acting on it. The object's free-body diagram is simply a sketch of the object *freed* from its surroundings showing *all* the (external) forces that *act* on it. The diagram focuses your attention on the object of interest and helps you identify *all* the external forces acting. For example:

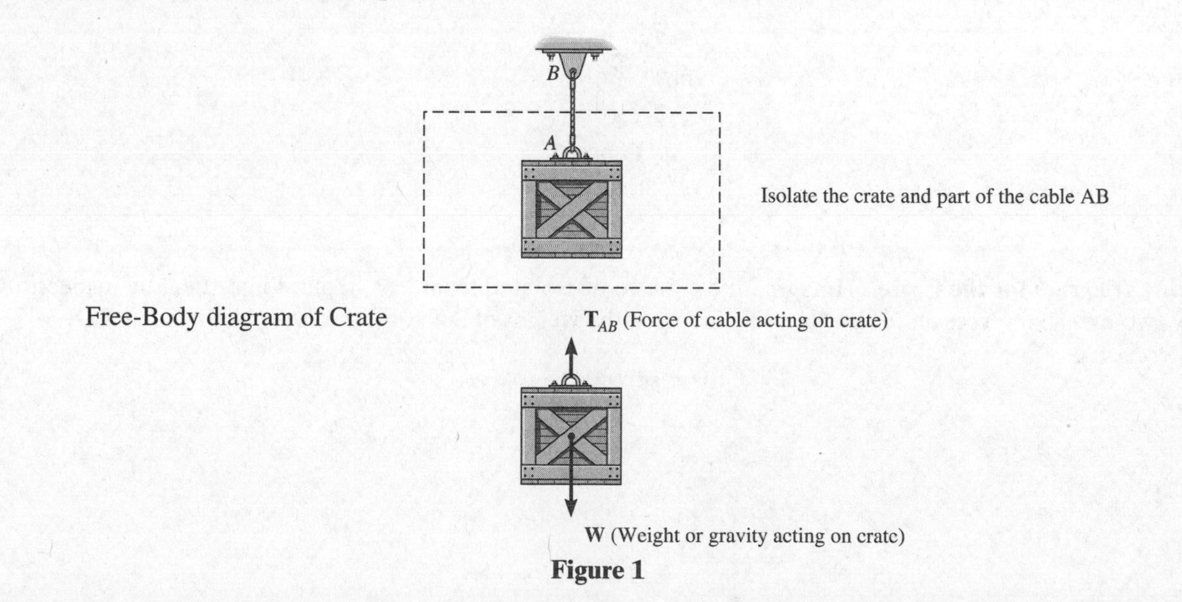

Figure 1

Note that once the crate is *separated* or *freed* from the system, forces which were previously internal to the system become external to the crate. For example, in Figure 1, such a force is the force of the cable AB *acting on the crate.*

Next, we present a formal procedure for drawing free-body diagrams for a particle.

2.1.1 Procedure for Drawing a Free-Body Diagram: Particle

1. *Identify the object you wish to isolate.* This choice is often dictated by the particular forces you wish to determine.
2. *Draw the outlined shape of the isolated object.* Imagine the object to be isolated or cut free from the system of which it is a part.
3. *Show all external forces acting on the isolated object.* Indicate on this sketch *all* the external forces that act on the object. These forces can be *active forces*, which tend to set the object in motion, or they can be *reactive forces* which are the result of the constraints or supports that prevent motion. This stage is crucial: it may help to trace around the object's boundary, carefully noting each external force acting on it. Don't forget to include the weight of the object (unless it is being intentionally neglected).
4. *Identify and label each external force acting on the (isolated) object.* The forces that are known should be labeled with their known magnitudes and directions. Use letters to represent the magnitudes and arrows to represent the directions of forces that are unknown.
5. *The direction of a force having an unknown magnitude can be assumed.*

EXAMPLE 2.1

The crate in Figure 2 has a weight of 20lb. Draw free-body diagrams of the crate, the cord BD and the ring at B. Assume that the cords and the ring at B have negligible mass.

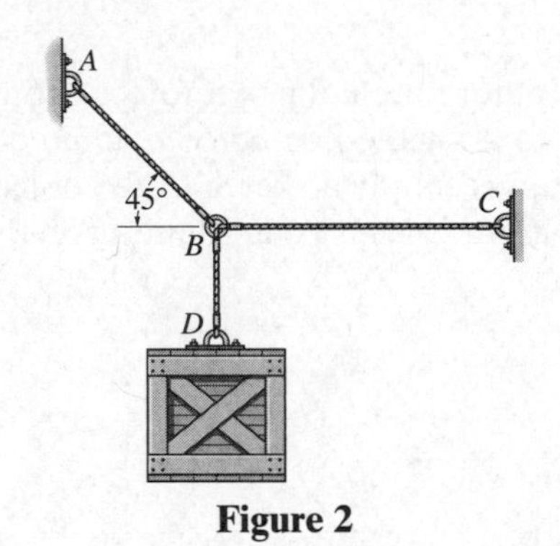

Figure 2

Solution

Free-Body Diagram for the Crate Imagine the crate to be isolated from its surroundings, then, by inspection, there are only two external forces *acting on the crate*, namely, the weight of 20lb and the force of the cord BD.

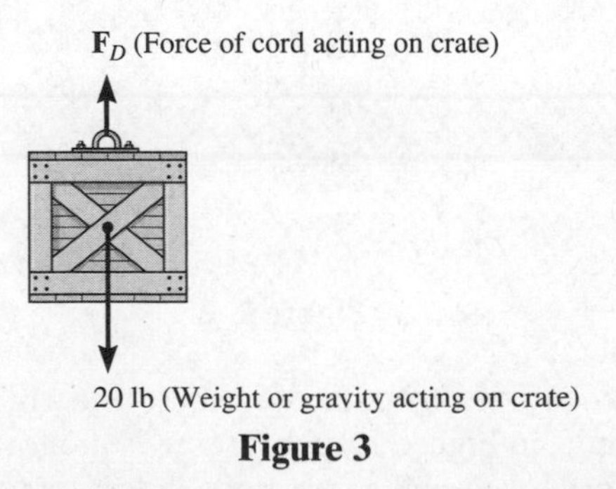

Figure 3

Free-Body Diagram for the Cord BD Imagine the cord to be isolated from its surroundings, then, by inspection, there are only two external forces *acting on the cord*, namely, the force of the crate F_D and the force F_B caused by the ring. These forces both tend to pull on the cord so that the cord is in *tension*. Notice that F_D shown in this free-body diagram (Figure 4) is equal and opposite to that shown in Figure 3 (a consequence of Newton's third law).

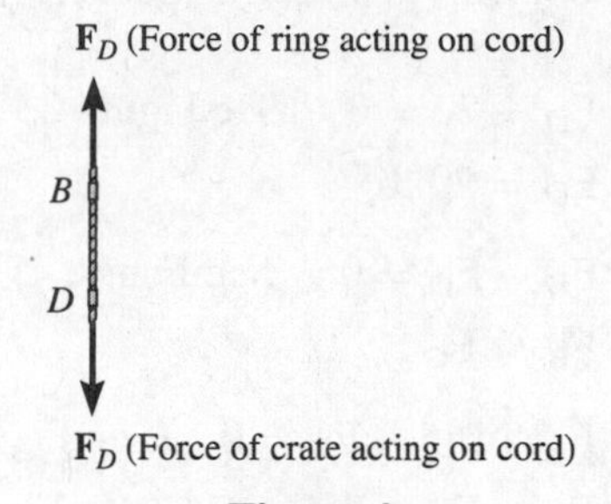

Figure 4

Free-Body Diagram for the ring at B Imagine the ring to be isolated from its surroundings, then, by inspection, there are actually three external forces acting on the ring, all caused by the attached cords. Notice that F_B shown in this free-body diagram (Figure 5) is equal and opposite to that shown in Figure 4 (a consequence of Newton's third law).

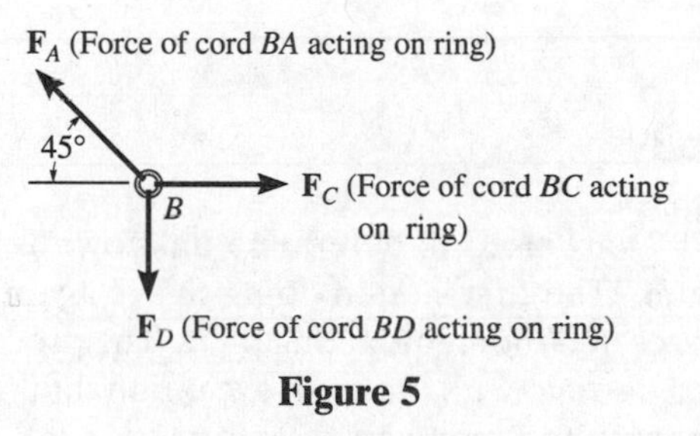

Figure 5

2.1.2 Using the Free-Body Diagram: Equilibrium

The free-body diagram is used to identify the unknown forces acting on the particle when applying the equilibrium equation (1.1) to the particle. The procedure for solving equilibrium problems for a particle once the free-body diagram for the particle is established, is therefore as follows:

1. *Establish* the x, y-axes in any suitable orientation.
2. Apply the equilibrium equation (1.1) in component form in each direction:

$$\sum F_x = 0 \text{ and } \sum F_y = 0 \qquad (2.1)$$

3. Components are positive if they are directed along a positive axis and negative if they are directed along a negative axis.
4. If more than two unknowns exist and the problem involves a spring, apply $F = ks$ to relate the magnitude of the spring force F to the deformation of the spring s (here, k is the spring constant).
5. If the solution yields a negative result, this indicates the sense of the force is the reverse of that shown/assumed on the free-body diagram.

EXAMPLE 2.2

In Example 2.1, the free-body diagrams established in Figures 3 - 5 give us a 'pictorial representation' of all the information we need to apply the equilibrium equations (2.1) to find the various unknown forces. In fact, taking the positive x-direction to be horizontal ($\rightarrow$ +) and the positive y-direction to be vertical ($\uparrow$ +), the equilibrium equations (2.1) when applied to each of the objects (regarded as particles) are:

For the Crate: $\uparrow + \sum F_y = 0$: $F_D - 20 = 0$ (See Figure 3)

$$F_D = 20 \text{ lb} \tag{2.2}$$

For the Cord BD: $\uparrow + \sum F_y = 0$: $F_B - F_D = 0$ (See Figure 4)

$$F_B = F_D \tag{2.3}$$

For the Ring: $\uparrow + \sum F_y = 0$: $F_A \sin 45 - F_B = 0$ (See Figure 5) (2.4)

$\longrightarrow + \sum F_x = 0$: $F_C - F_A \cos 45 = 0$ (See Figure 5) (2.5)

Equations (2.2)–(2.5) are now 4 equations which can be solved for the 4 unknowns F_A, F_B, F_C and F_D. That is: $F_B = 20$ lb; $F_D = 20$ lb, $F_A = 28.28$, $F_C = 20$. The directions of each of these forces is shown in the free-body diagrams above (Figures 3–5). ◀

2.2 Free-Body Diagram: Rigid Body

The equilibrium equations (1.2) and (1.3) are used to determine unknown forces and moments acting *on an object* (modeled as a rigid body) in equilibrium. The first step in doing this is again to draw the *free-body diagram* of the object to identify *all of* the external forces and moments acting on it. The procedure for drawing a free-body diagram in this case is much the same as that for a particle with the main exception that now, because the object has 'size/shape,' it can support also external couple moments and moments of external forces.

2.2.1 Procedure for Drawing a Free-Body Diagram: Rigid Body

1. Imagine the body to be isolated or 'cut free' from its constraints and connections and sketch its outlined shape.
2. Identify all the external forces and couple moments that act on the body. Those generally encountered are:
   (a) Applied loadings
   (b) Reactions occurring at the supports or at points of contact with other bodies (See Table 2.1)
   (c) The weight of the body (applied at the body's center of gravity G)
3. The forces and couple moments that are known should be labeled with their proper magnitudes and directions. Letters are used to represent the magnitudes and direction angles of forces and couple moments that are *unknown*. Establish an x, y-coordinate system so that these unknowns e.g. A_x, B_y etc. can be identified. Indicate the dimensions of the body necessary for computing the moments of external forces. In particular, if a force or couple moment has a known line of action but unknown magnitude, the arrowhead which defines the sense of the vector can be assumed. The correctness of the assumed sense will become apparent after solving the equilibrium equations for the unknown magnitude. By definition, the magnitude of a vector is *always positive*, so that if the solution yields a *negative* scalar, the minus *sign* indicates that the vector's sense is *opposite* to that which was originally assumed.

Table 2.1. Supports for Rigid Bodies Subjected to Two-Dimensional Force Systems

	Types of Connection	Reaction	Number of Unknowns
(1)	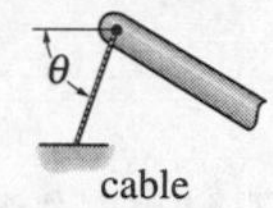 cable (θ)	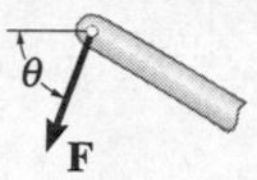 θ, $\mathbf{F}$	One unknown. The reaction is a tension force which acts away from the member in the direction of the cable.
(2)	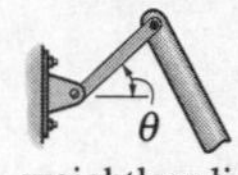 weightless link (θ)	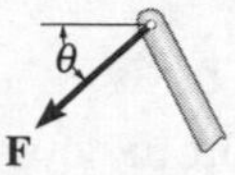 θ, $\mathbf{F}$ or θ, $\mathbf{F}$ 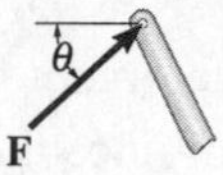	One unknown. The reaction is a force which acts along the axis of the link.
(3)	 roller (θ)	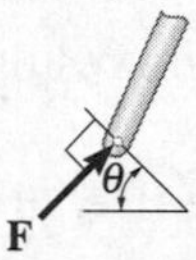 θ, $\mathbf{F}$	One unknown. The reaction is a force which acts perpendicular to the surface at the point of contact.
(4)	 roller or pin in confined smooth slot (θ)	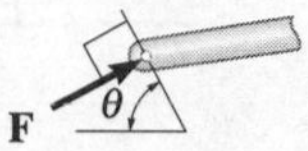 $\mathbf{F}$, θ or $\mathbf{F}$, θ 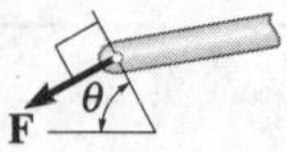	One unknown. The reaction is a force which acts perpendicular to the slot.
(5)	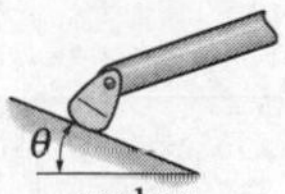 rocker (θ)	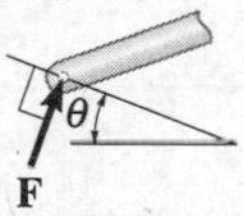 θ, $\mathbf{F}$	One unknown. The reaction is a force which acts perpendicular to the surface at the point of contact.
(6)	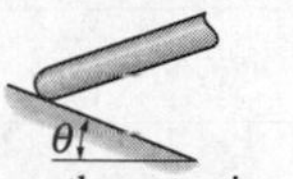 smooth contacting surface (θ)	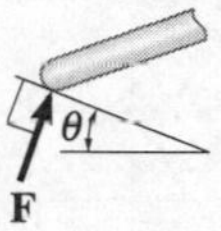 θ, $\mathbf{F}$	One unknown. The reaction is a force which acts perpendicular to the surface at the point of contact.
(7)	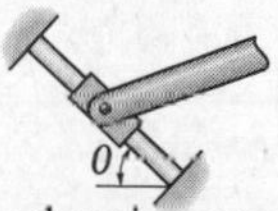 member pin connected to collar on smooth rod (θ)	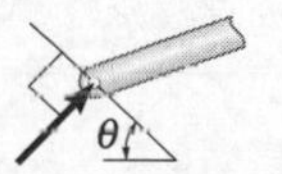 θ or 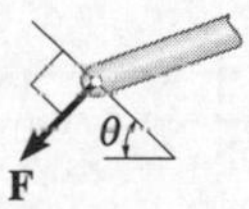 θ, $\mathbf{F}$	One unknown. The reaction is a force which acts perpendicular to the rod.
(8)	 smooth pin or hinge (θ)	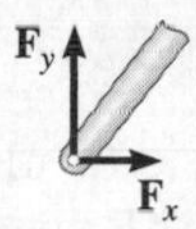 $\mathbf{F}_y$, $\mathbf{F}_x$ or $\mathbf{F}$, ϕ 	Two unknowns. The reactions are two components of force, or the magnitude and direction ϕ of the resultant force. Note that ϕ and θ are not necessarily equal [usually not, unless the rod shown is a link as in (2)].
(9)	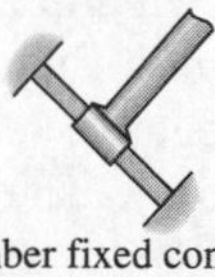 member fixed connected to collar on smooth rod	 $\mathbf{M}$, $\mathbf{F}$	Two unknowns. The reactions are the couple moment and the force which acts perpendicular to the rod.
(10)	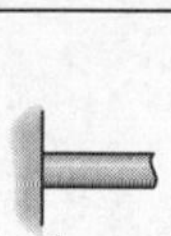 fixed support	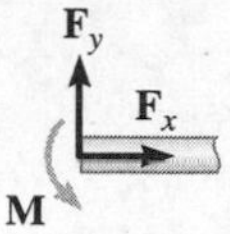 $\mathbf{F}_y$, $\mathbf{F}_x$, $\mathbf{M}$ or $\mathbf{F}$, ϕ, $\mathbf{M}$ 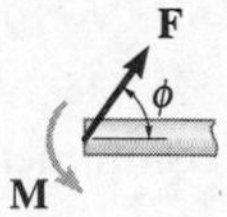	Three unknowns. The reactions are the couple moment and the two force components, or the couple moment and the magnitude and direction ϕ of the resultant force.

2.2.2 *Important Points*

- No equilibrium problem should be solved without first drawing the free-body diagram, so as to account for all the external forces and moments that act on the body.
- If a support *prevents translation* of a body in a particular direction, then the support exerts a force on the body in that direction
- If *rotation is prevented* then the support exerts a couple moment on the body
- Internal forces are never shown on the free-body diagram since they occur in equal but opposite collinear pairs and therefore cancel each other out.
- The weight of a body is an external force and its effect is shown as a single resultant force acting through the body's center of gravity G.
- *Couple moments* can be placed anywhere on the free-body diagram since they are *free vectors*. Forces can act at any point along their lines of action since they are *sliding vectors*.

EXAMPLE 2.3

Draw the free-body diagram of the beam, which is pin-connected at A and rocker-supported at B. Neglect the weight of the beam.

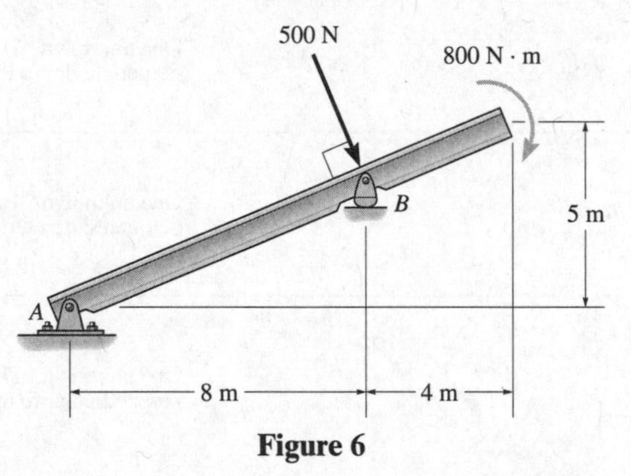

Figure 6

Solution

The free-body diagram of the beam is shown in Figure 7. From Table 2.1, since the support at A is a pin-connection, there are two reactions acting on the beam at A denoted by $\mathbf{A}_x$ and $\mathbf{A}_y$. In addition, there is one reaction acting on the beam at the rocker support at B. We denote this reaction by the force **F** which acts perpendicular to the surface at B, the point of contact (see Table 2.1). The magnitudes of these vectors are unknown and their sense has been assumed (the correctness of the assumed sense will become apparent after solving the equilibrium equations for the unknown magnitude i.e. if application of the equilibrium equations to the beam yields a negative result for **F**, this indicates the sense of the force is the reverse of that shown/assumed on the free-body diagram). The weight of the beam has been neglected. ◀

2.2.3 *Using the Free-Body Diagram: Equilibrium*

The equilibrium equations (1.2) and (1.3) can be written in component form as:

$$\sum F_x = 0, \tag{2.6}$$

$$\sum F_y = 0, \tag{2.7}$$

$$\sum M_O = 0. \tag{2.8}$$

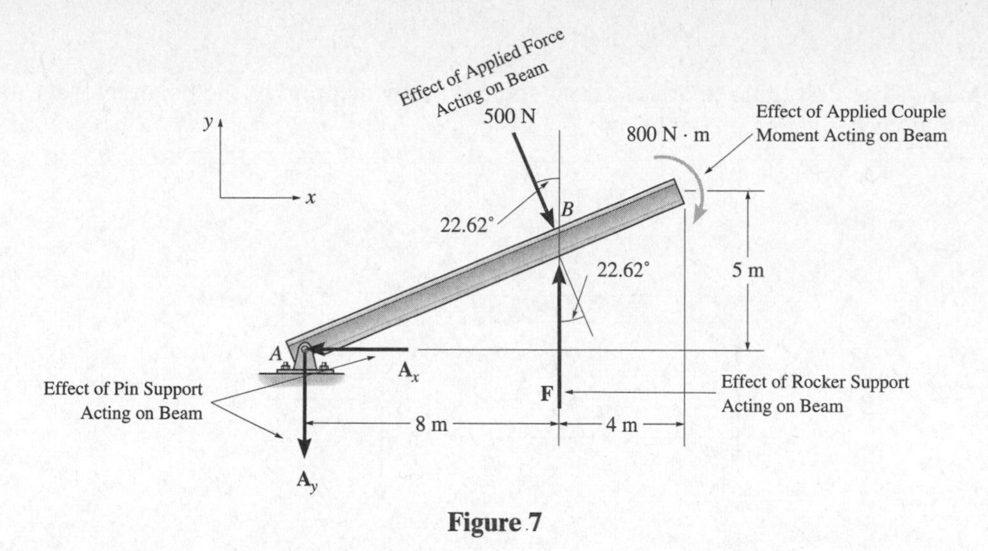

Figure 7

Here, $\sum F_x$ and $\sum F_y$ represent, respectively, the algebraic sums of the x and y components of all the external forces acting on the body and $\sum M_O$ represents the algebraic sum of the couple moments and the moments of all the external force components about an axis perpendicular to the x-y plane and passing through the arbitrary point O, which may lie either on or off the body. The procedure for solving equilibrium problems for a rigid body once the free-body diagram for the body is established, is as follows:

- Apply the moment equation of equilibrium (2.8), about a point (O) that lies at the intersection of the lines of action of two unknown forces. In this way, the moments of these unknowns are zero about O and a direct solution for the third unknown can be determined.
- When applying the force equilibrium equations (2.6) and (2.7), orient the x and y-axes along lines that will provide the simplest resolution of the forces into their x and y components.
- If the solution of the equilibrium equations yields a negative scalar for a force or couple moment magnitude, this indicates that the sense is opposite to that which was assumed on the free-body diagram.

Example 2.4

A force of magnitude 150 lb acts on the end of the beam as shown. Find the magnitude and direction of the reaction at pin A and the tension in the cable.

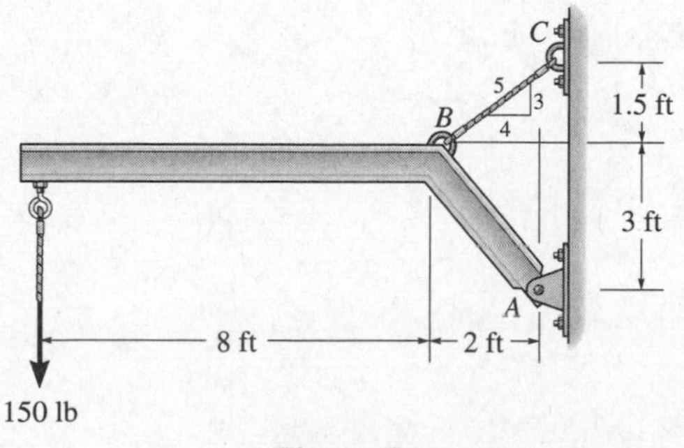

Figure 8

Solution

Free-Body Diagram The first thing to do is to draw the free-body diagram of the beam in order to identify all the external forces and moments acting on the beam.

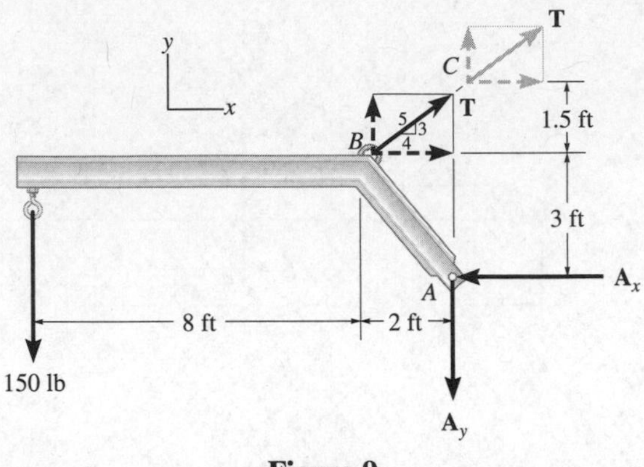

Figure 9

Equations of Equilibrium The free-body diagram of the beam suggests we can sum moments about the point A to eliminate the moment contribution of the reaction forces $\mathbf{A}_x$ and $\mathbf{A}_y$ acting on the beam. This will allow us to obtain a direct solution for the third unknown i.e. the cable tension T. Taking counterclockwise as positive when computing moments, we have:

$$+\circlearrowleft \sum M_A = 0: \quad -(3/5)T(2\text{ ft}) - (4/5)T(3\text{ ft}) + 150\text{ lb}(10\text{ ft}) = 0$$

$$-3.6T + 150\text{ lb}(10\text{ ft}) = 0$$

$$\underline{T = 416.7\text{ lb}} \qquad \textbf{Ans.}$$

Summing forces to obtain A_x and A_y, using the result for T, we have

$$\xrightarrow{} + \sum F_x = 0: \quad -A_x + (4/5)(416.7\text{ lb}) = 0$$

$$\mathbf{A}_x = 333.3\text{ lb} \longleftarrow$$

$$\uparrow + \sum F_y = 0: \quad (3/5)(416.7\text{ lb}) - 150\text{ lb} - A_y = 0$$

$$\mathbf{A}_y = 100\text{ lb} \downarrow$$

Thus, the reaction force $\mathbf{F}_A$ at pin A has magnitude F_A given by:

$$F_A = \sqrt{[(333.3\text{ lb})^2 + (100\text{ lb})^2]} = 348.0\text{ lb}$$

and direction given by

$$\theta = \tan^{-1}[(-100\text{ lb})/(-333.3\text{ lb})] = 196.7^\circ$$

counterclockwise from the positive x-axis or  ◀

3

Problems

3.1 Free-Body Diagrams in Particle Equilibrium

Problem 3.1

The sling is used to support a drum having a weight of 900 lb. Draw a free-body diagram for the knot at A. Take $\theta = 20°$.

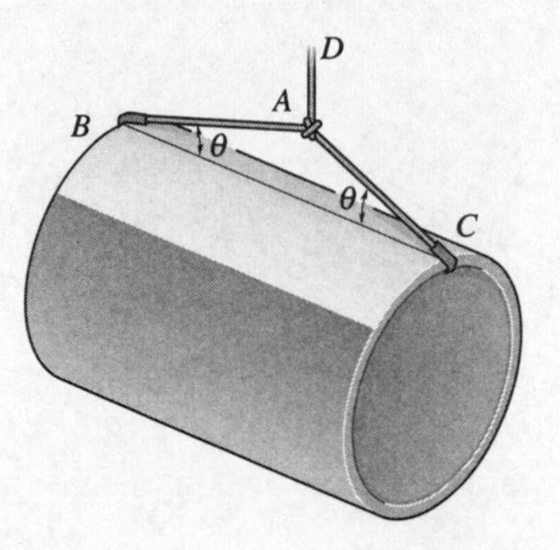

Solution

1. The knot at A has *negligible size* so that it can be modelled as a particle.
2. Imagine the knot at A to be separated or detached from the system.
3. The (detached) knot at A is subjected to three *external* forces. They are caused by:

   i. **ii.**

   iii.
4. Draw the free-body diagram of the (detached) knot showing all these forces labeled with their magnitudes and directions.

Problem 3.1

The sling is used to support a drum having a weight of 900 lb. Draw a free-body diagram for the knot at A. Take $\theta = 20°$.

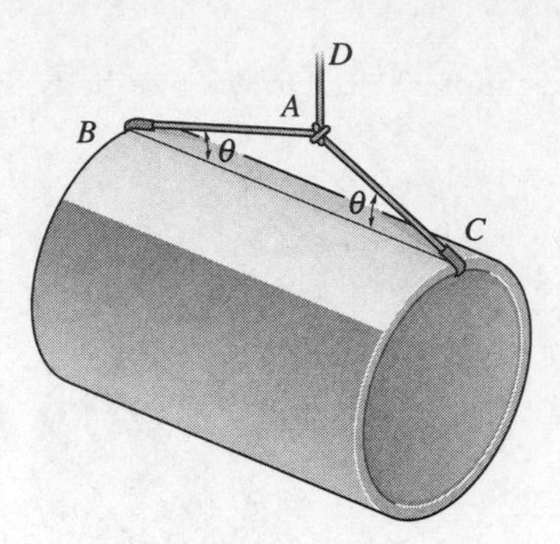

Solution

1. The knot at A has *negligible size* so that it can be modelled as a particle.
2. Imagine the knot at A to be separated or detached from the system.
3. The (detached) knot at A is subjected to three *external* forces. They are caused by:
   - i. **CORD** AB
   - ii. **CORD** AC
   - iii. **CORD** AD *(weight of drum)*
4. Draw the free-body diagram of the (detached) knot showing all these forces labeled with their magnitudes and directions.

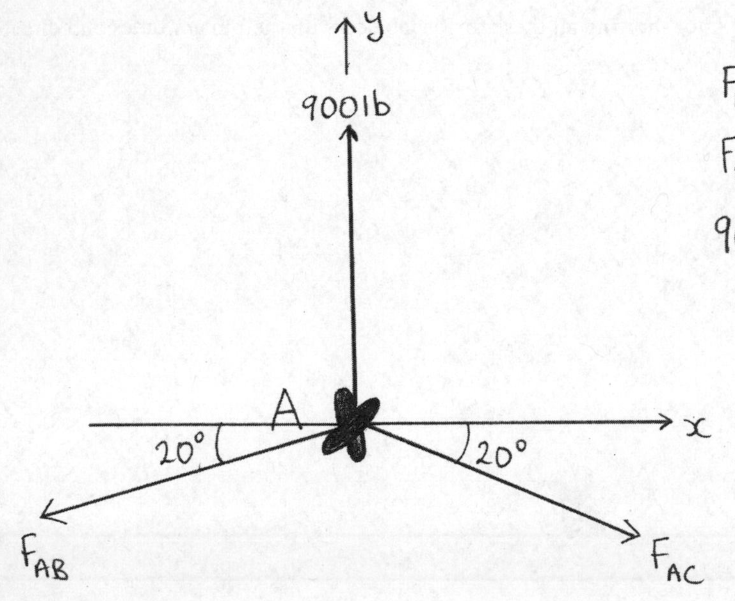

F_{AB} force of cord AB on knot
F_{AC} force of cord AC on knot
900lb force of gravity on knot
(force of cord AD on knot)

Problem 3.2

The spring ABC has a stiffness of 500 N/m and an unstretched length of 6 m. A horizontal force **F** is applied to the cord which is attached to the *small* pulley B so that the displacement of the pulley from the wall is $d = 1.5$ m. Draw a free-body diagram for the small pulley B.

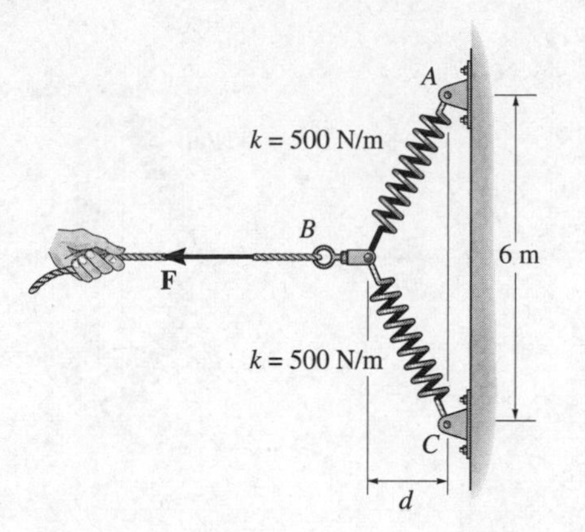

Solution

1. The pulley B has *negligible size* so that it can be modelled as a particle.
2. Imagine the pulley B to be separated or detached from the system.
3. The (detached) pulley B is subjected to three *external* forces. They are caused by:

   i.

   ii.

   iii.

4. Draw the free-body diagram of the (detached) pulley showing all these forces labeled with their magnitudes and directions. You should also include any other available information e.g. lengths, angles etc. — which will help when formulating the equilibrium equations for the pulley.

Problem 3.2

The spring ABC has a stiffness of 500 N/m and an unstretched length of 6 m. A horizontal force **F** is applied to the cord which is attached to the *small* pulley B so that the displacement of the pulley from the wall is $d = 1.5$ m. Draw a free-body diagram for the small pulley B.

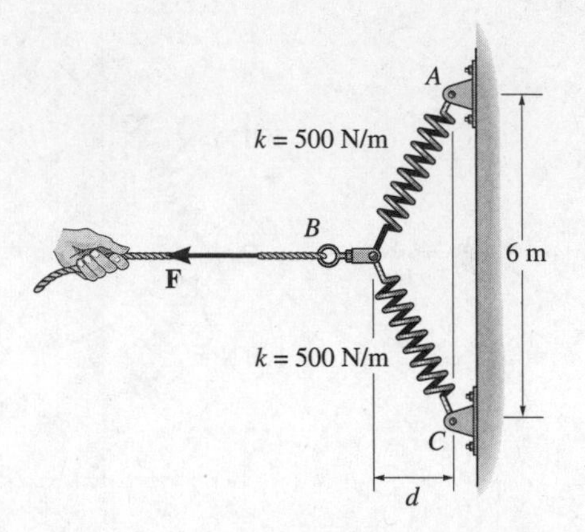

Solution

1. The pulley B has *negligible size* so that it can be modelled as a particle.
2. Imagine the pulley B to be separated or detached from the system.
3. The (detached) pulley B is subjected to three *external* forces. They are caused by:
   i. **Force F**
   ii. **Spring** AB
   iii. **Spring** BC
4. Draw the free-body diagram of the (detached) pulley showing all these forces labeled with their magnitudes and directions. You should also include any other available information e.g. lengths, angles etc. — which will help when formulating the equilibrium equations for the pulley.

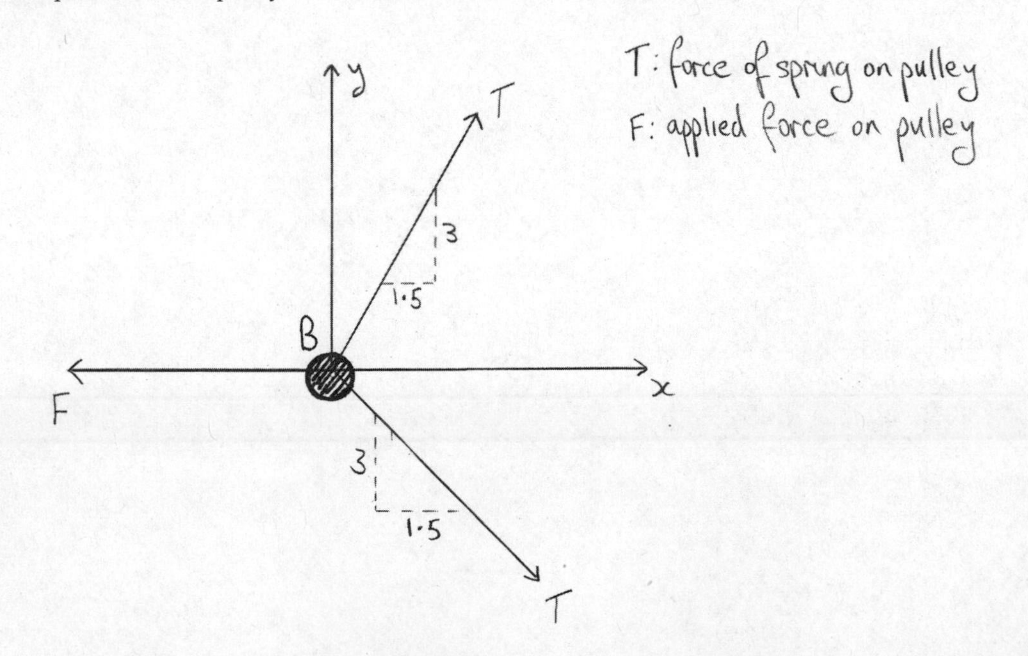

Problem 3.3

The 2-kg block is held in equilibrium by the system of springs. Draw a free-body diagram for the ring at A.

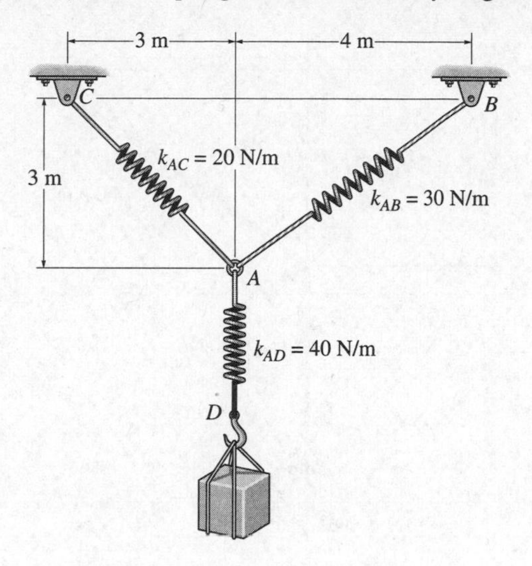

Solution

1. The ring at A has *negligible size* so that it can be modelled as a particle.
2. Imagine the ring at A to be separated or detached from the system.
3. The (detached) ring at A is subjected to three *external* forces. They are caused by:

   i.

   ii.

   iii.

4. Draw the free-body diagram of the (detached) ring showing all these forces labeled with their magnitudes and directions. You should also include any other available information e.g. lengths, angles etc. — which will help when formulating the equilibrium equations for the ring.

Problem 3.3

The 2-kg block is held in equilibrium by the system of springs. Draw a free-body diagram for the ring at A.

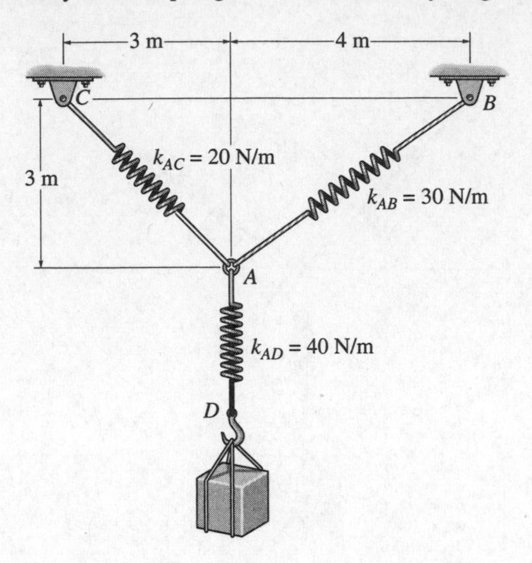

Solution

1. The ring at A has *negligible size* so that it can be modelled as a particle.
2. Imagine the ring at A to be separated or detached from the system.
3. The (detached) ring at A is subjected to three *external* forces. They are caused by:

   i. Spring AD *(weight of block)* **ii. Spring** AC

   iii. Spring AB

4. Draw the free-body diagram of the (detached) ring showing all these forces labeled with their magnitudes and directions. You should also include any other available information e.g. lengths, angles etc. — which will help when formulating the equilibrium equations for the ring.

F_{AC} F_{AB} y x A $2(9.81)$N

F_{AC}: force of spring AC on ring
F_{AB}: force of spring AB on ring
$2(9.81)$: weight of block on ring
(force of spring AD on ring)

Problem 3.4

The motor at B winds up the cord attached to the 65-lb crate with a constant speed. The force in cord CD supports the pulley C and the angle θ represents the equilibrium state. Draw the free-body diagram of the pulley C. *Neglect the size of the pulley.*

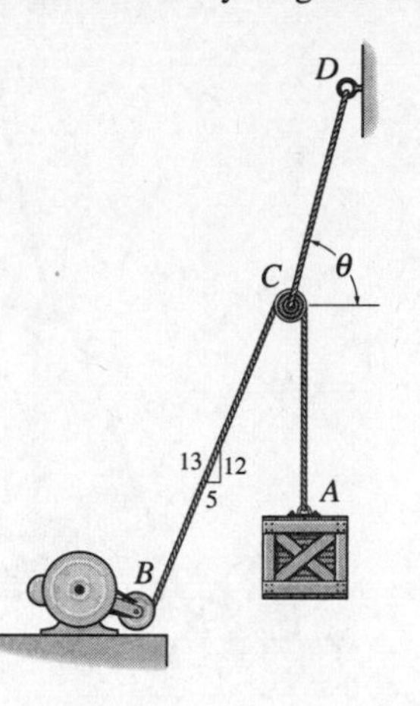

Solution

1. The pulley C has *negligible size* so that it can be modelled as a particle.
2. Imagine the pulley C to be separated or detached from the system.
3. The (detached) pulley C is subjected to three *external* forces. They are caused by:

   i.

   ii.

   iii.

4. Draw the free-body diagram of the (detached) pulley showing all these forces labeled with their magnitudes and directions. You should also include any other available information e.g. lengths, angles etc. — which will help when formulating the equilibrium equations for the pulley.

Problem 3.4

The motor at B winds up the cord attached to the 65-lb crate with a constant speed. The force in cord CD supports the pulley C and the angle θ represents the equilibrium state. Draw the free-body diagram of the pulley C. *Neglect the size of the pulley.*

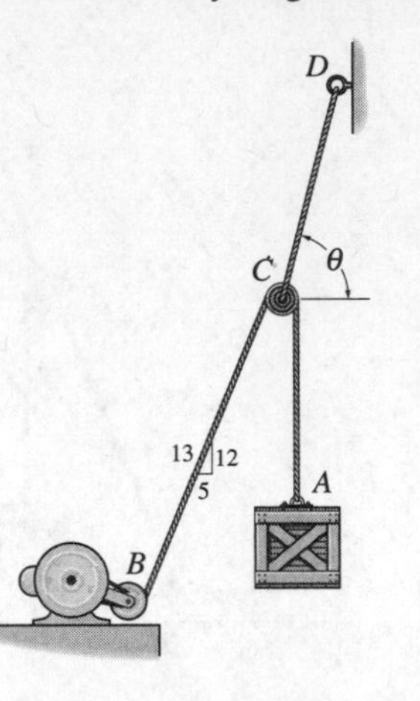

Solution

1. The pulley C has *negligible size* so that it can be modelled as a particle.
2. Imagine the pulley C to be separated or detached from the system.
3. The (detached) pulley C is subjected to three *external* forces. They are caused by:

   i. CORD CD

   ii. CORD CB

   iii. CORD CA *(weight of crate)*

4. Draw the free-body diagram of the (detached) pulley showing all these forces labeled with their magnitudes and directions. You should also include any other available information e.g. lengths, angles etc. — which will help when formulating the equilibrium equations for the pulley.

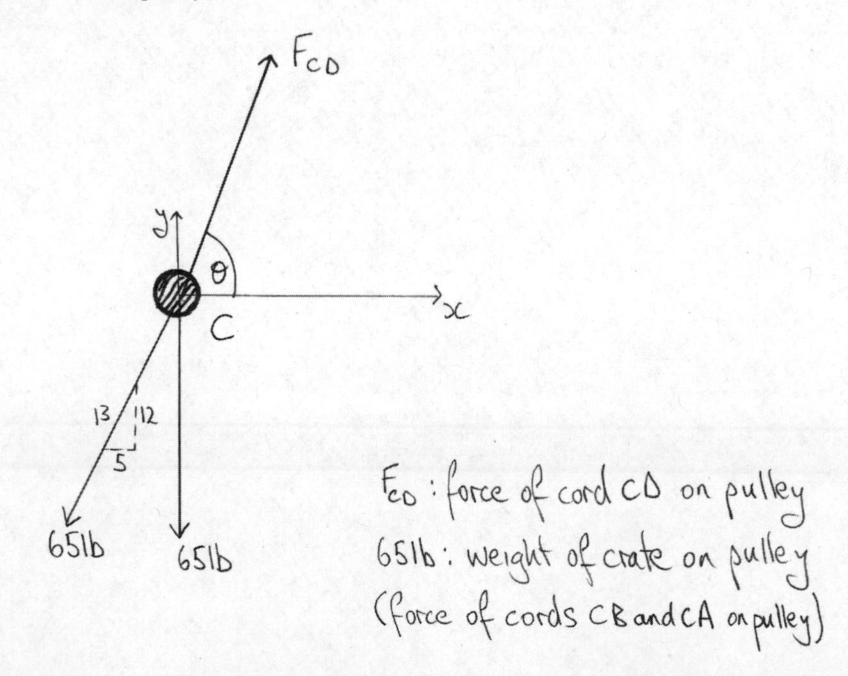

Problem 3.5

The following system is held in equilibrium by the mass supported at A and the angle θ of the connecting cord. Draw the free-body diagram for the connecting knot D.

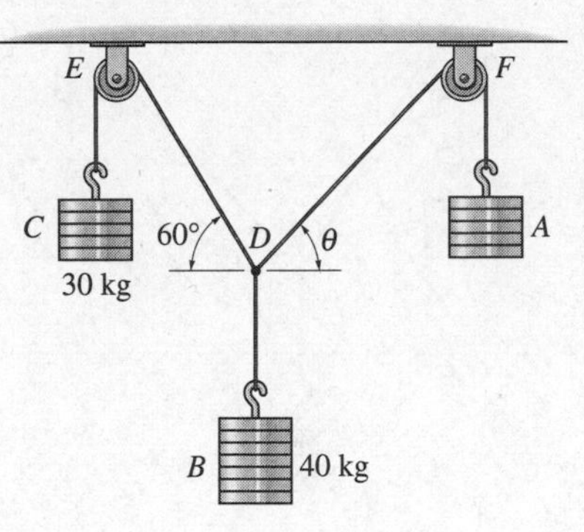

Solution

1. The knot D has *negligible size* so that it can be modelled as a particle.
2. Imagine the knot D to be separated or detached from the system.
3. The (detached) knot D is subjected to three *external* forces. They are caused by:

   i. **ii.**

   iii.

4. Draw the free-body diagram of the (detached) knot showing all these forces labeled with their magnitudes and directions. You should also include any other available information e.g. lengths, angles etc. — which will help when formulating the equilibrium equations for the knot.

Problem 3.5

The following system is held in equilibrium by the mass supported at A and the angle θ of the connecting cord. Draw the free-body diagram for the connecting knot D.

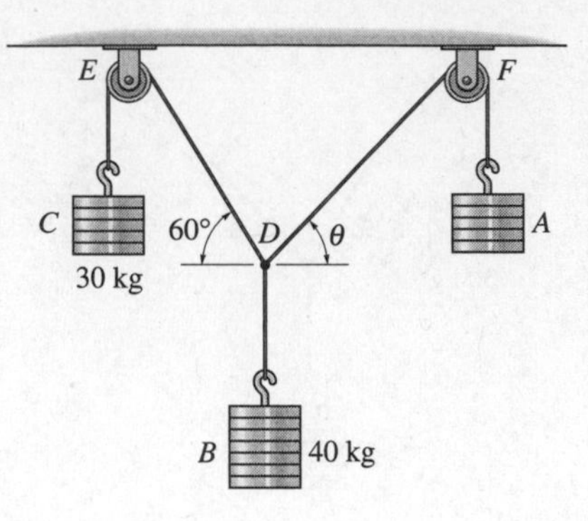

Solution

1. The knot D has *negligible size* so that it can be modelled as a particle.
2. Imagine the knot D to be separated or detached from the system.
3. The (detached) knot D is subjected to three *external* forces. They are caused by:
   i. **CORD** DE *(weight of C)*
   ii. **CORD** DF *(weight of A)*
   iii. **CORD** DB *(weight of B)*
4. Draw the free-body diagram of the (detached) knot showing all these forces labeled with their magnitudes and directions. You should also include any other available information e.g. lengths, angles etc. — which will help when formulating the equilibrium equations for the knot.

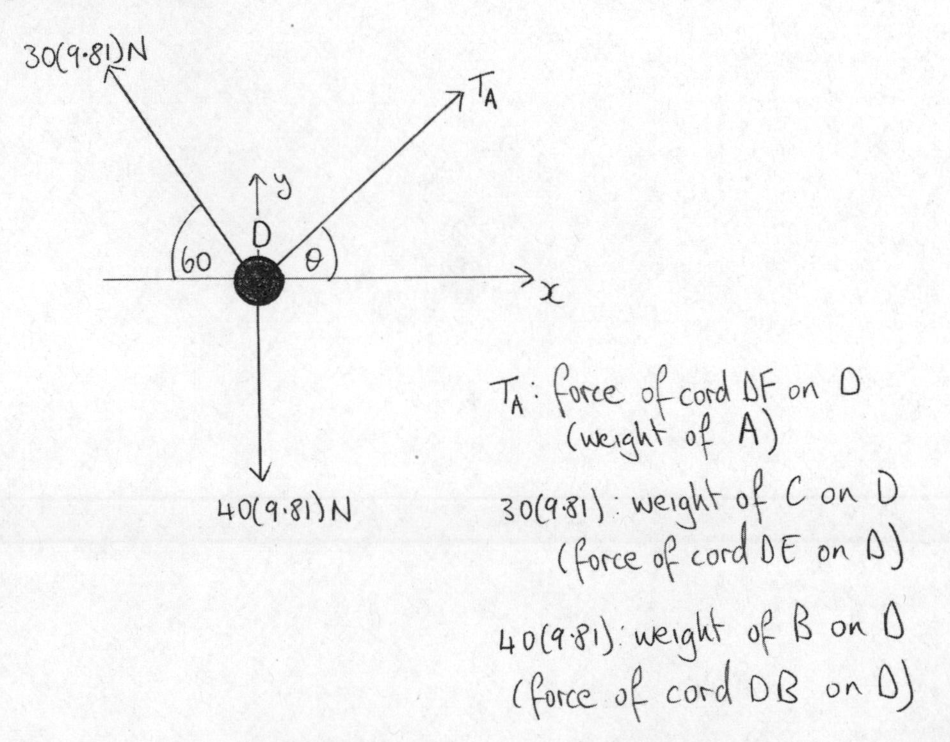

Problem 3.6

The 500 lb crate is hoisted using the ropes AB and AC. Each rope can withstand a maximum tension of 2500 lb before it breaks. Rope AB always remains horizontal. Draw the free-body diagram for the ring at A and determine the smallest angle θ to which the crate can be hoisted.

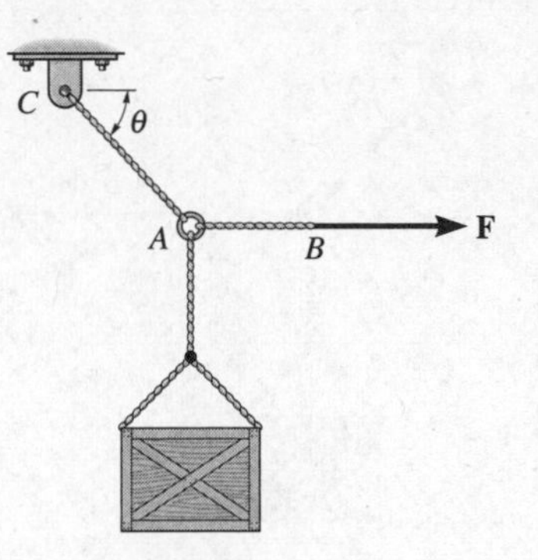

Solution

1. The ring at A has *negligible size* so that it can be modelled as a particle.
2. Imagine the ring at A to be separated or detached from the system.
3. The (detached) ring A is subjected to three *external* forces. They are caused by:

   i. **ii.**

   iii.

4. Draw the free-body diagram of the (detached) ring showing all these forces labeled with their magnitudes and directions.

5. Establish an xy-axes system on the free-body diagram and write down the equilibrium equations in each of the x and y-directions

   $+\uparrow \sum F_y = 0:$

   $\underset{\rightarrow}{+} \sum F_x = 0:$

6. Solve for the angle θ:

Problem 3.6

The 500 lb crate is hoisted using the ropes AB and AC. Each rope can withstand a maximum tension of 2500 lb before it breaks. Rope AB always remains horizontal. Draw the free-body diagram for the ring at A and determine the smallest angle θ to which the crate can be hoisted.

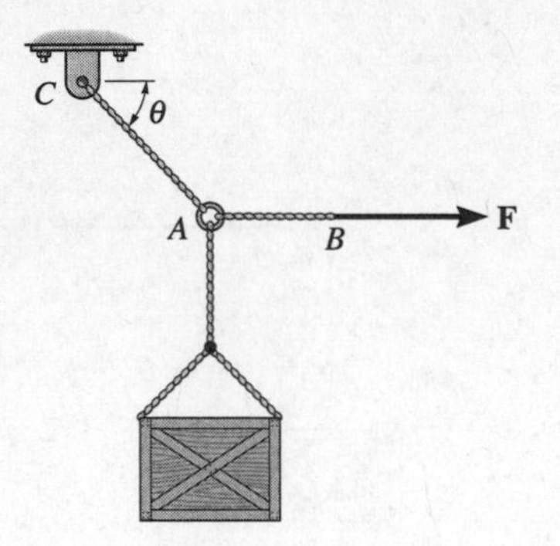

Solution

1. The ring at A has *negligible size* so that it can be modelled as a particle.
2. Imagine the ring at A to be separated or detached from the system.
3. The (detached) ring A is subjected to three *external* forces. They are caused by:

   i. **CORD** AC ii. **CORD** AB

   iii. **CORD** AD *(weight of crate)*
4. Draw the free-body diagram of the (detached) ring showing all these forces labeled with their magnitudes and directions.

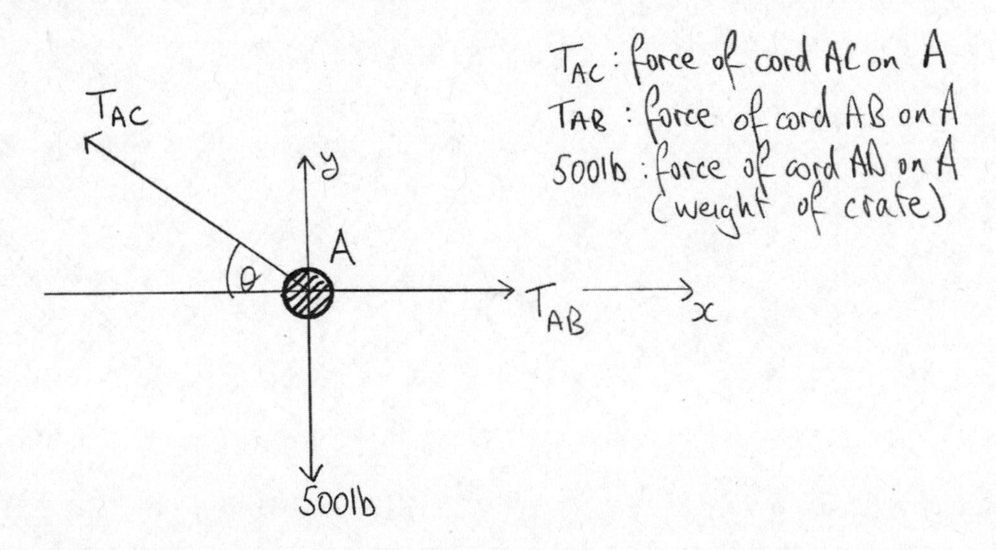

5. Establish an xy-axes system on the free-body diagram and write down the equilibrium equations in each of the x and y-directions

$$+\uparrow \sum F_y = 0: \; T_{AC} \sin\theta - 500 = 0$$

$$\underset{\rightarrow}{+} \sum F_x = 0: \; T_{AB} - T_{AC} \cos\theta = 0$$

6. Solve for the angle θ:

Assume $T_{AC} = 2500 \text{ lb} \Rightarrow \theta = 11.54^\circ$ and $T_{AB} = 2449.49 \text{ lb} < 2500 \text{ lb}$ (O.K!) **Ans.**

Problem 3.7

The block has a weight of 20 lb and is being hoisted at uniform velocity. The system is held in equilibrium at angle θ by the appropriate force in each cord. Draw the free-body diagram for the *small* pulley.

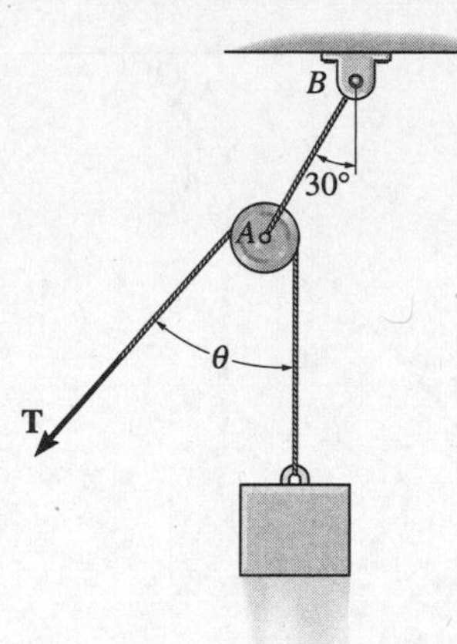

Solution

1. The pulley has *negligible size* so that it can be modelled as a particle.
2. Imagine the pulley to be separated or detached from the system.
3. The (detached) pulley is subjected to three *external* forces. They are caused by:
   i. **ii.**
   iii.
4. Draw the free-body diagram of the (detached) pulley showing all these forces labeled with their magnitudes and directions. You should also include any other available information e.g. lengths, angles etc. — which will help when formulating the equilibrium equations for the knot.

Problem 3.7

The block has a weight of 20 lb and is being hoisted at uniform velocity. The system is held in equilibrium at angle θ by the appropriate force in each cord. Draw the free-body diagram for the *small* pulley.

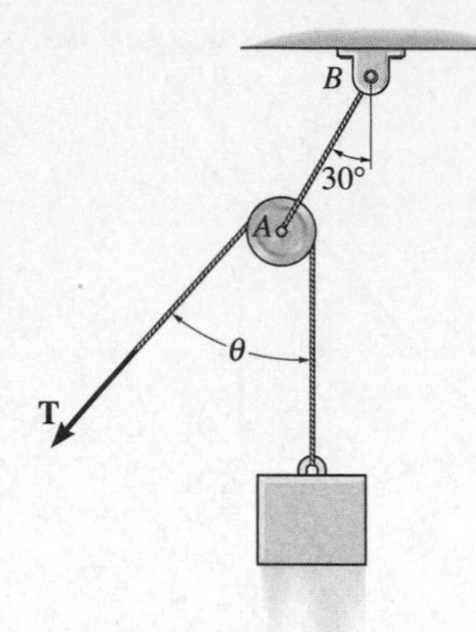

Solution

1. The pulley has *negligible size* so that it can be modelled as a particle.
2. Imagine the pulley to be separated or detached from the system.
3. The (detached) pulley is subjected to three *external* forces. They are caused by:
   i. Cord AB **ii. Force T**
   iii. Weight of block
4. Draw the free-body diagram of the (detached) pulley showing all these forces labeled with their magnitudes and directions. You should also include any other available information e.g. lengths, angles etc. — which will help when formulating the equilibrium equations for the knot.

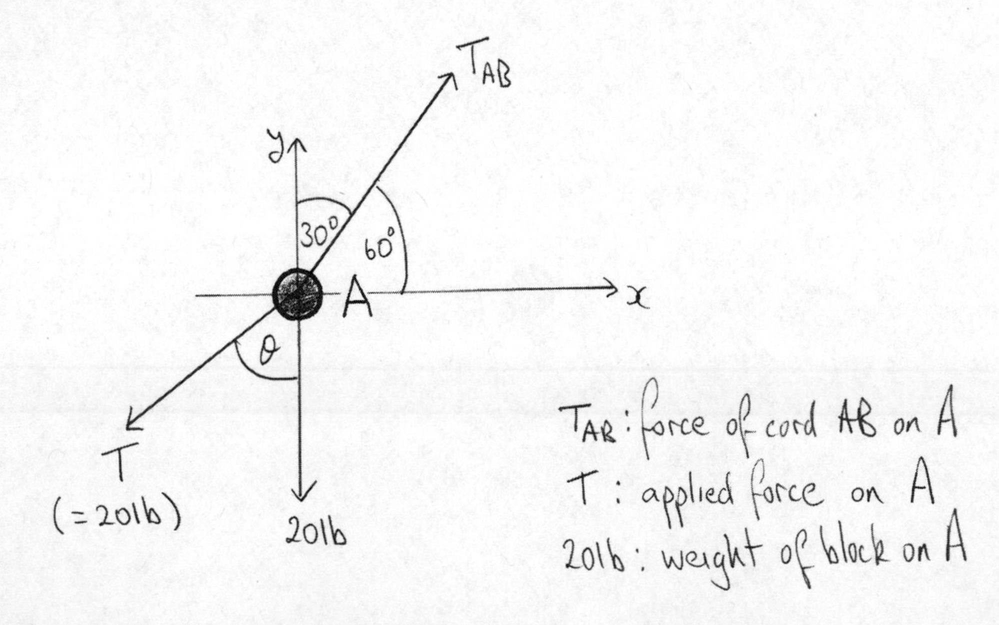

Problem 3.8

Blocks D and F weigh 5 lb each and block E weighs 8 lb. The system is in equilibrium at a given sag s. Draw the free-body diagram for the connecting ring at A and find s. Neglect the size of the pulleys.

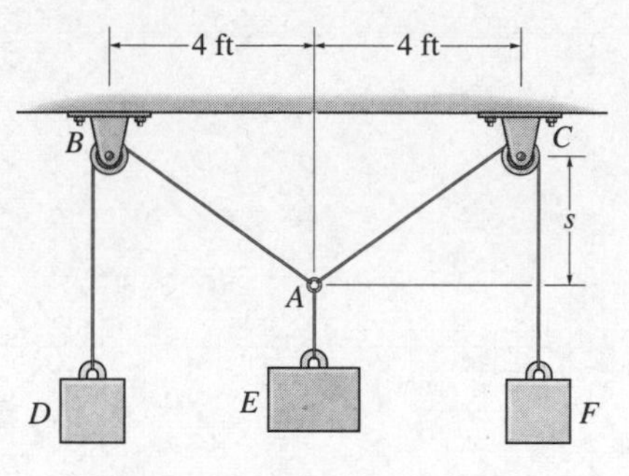

Solution

1. The ring at A has *negligible size* so that it can be modelled as a particle.
2. Imagine the ring to be separated or detached from the system.
3. The (detached) ring is subjected to three *external* forces. They are caused by:

   i. **ii.**

   iii.

4. Draw the free-body diagram of the (detached) ring showing all these forces labeled with their magnitudes and directions. Include also any other information which may help when formulating the equilibrium equations for the ring.

A

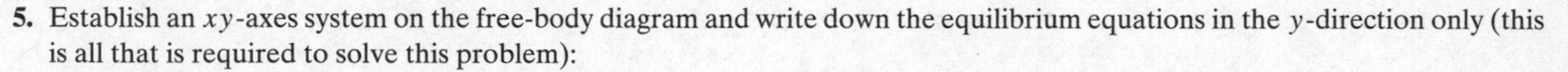

5. Establish an xy-axes system on the free-body diagram and write down the equilibrium equations in the y-direction only (this is all that is required to solve this problem):

   $+\uparrow \sum F_y = 0:$

6. Solve for the sag s:

Problem 3.8

Blocks D and F weigh 5 lb each and block E weighs 8 lb. The system is in equilibrium at a given sag s. Draw the free-body diagram for the connecting ring at A and find s. Neglect the size of the pulleys.

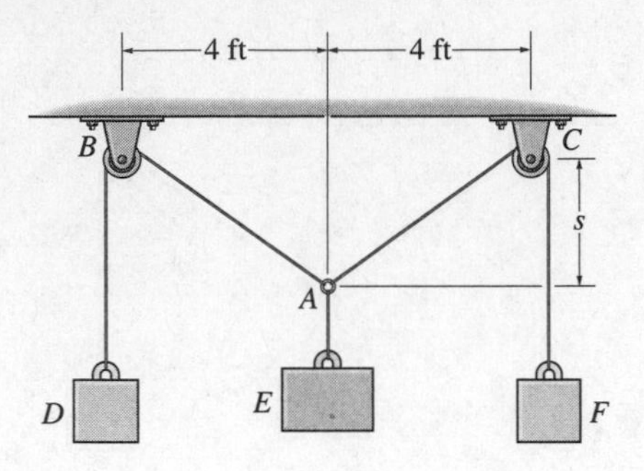

Solution

1. The ring at A has *negligible size* so that it can be modelled as a particle.
2. Imagine the ring to be separated or detached from the system.
3. The (detached) ring is subjected to three *external* forces. They are caused by:
   i. **CORD** AB *(weight of D)* ii. **CORD** AC *(weight of F)*
   iii. **CORD** AE *(weight of E)*
4. Draw the free-body diagram of the (detached) ring showing all these forces labeled with their magnitudes and directions. Include also any other information which may help when formulating the equilibrium equations for the ring.

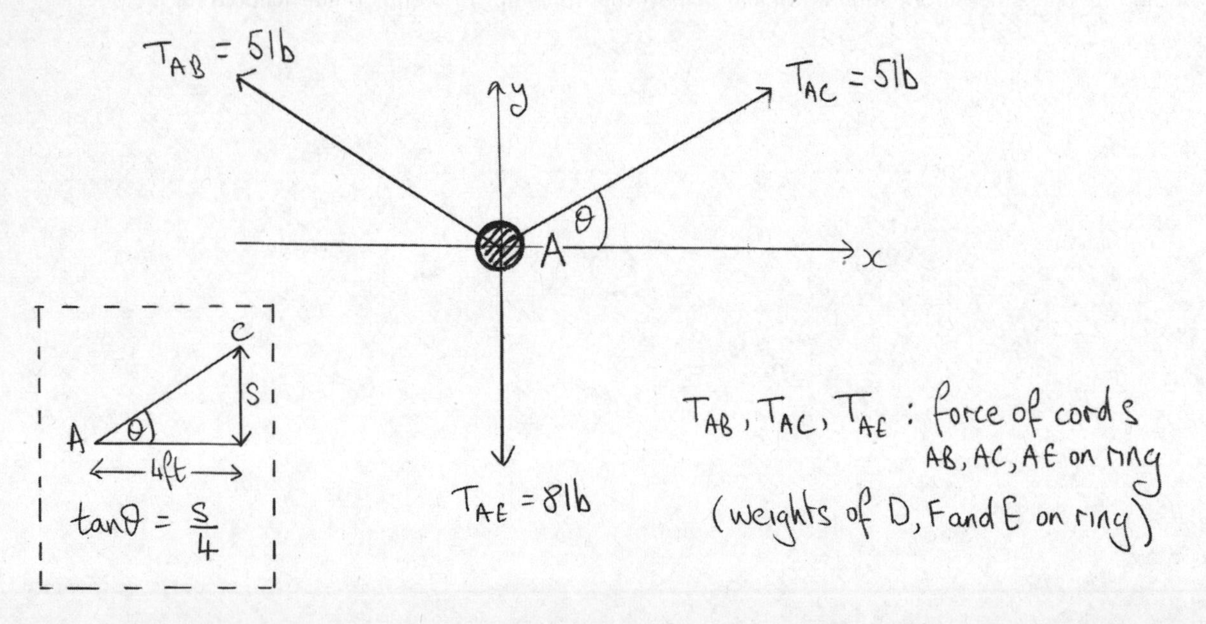

5. Establish an xy-axes system on the free-body diagram and write down the equilibrium equations in the y-direction only (this is all that is required to solve this problem):

$$+\uparrow \sum F_y = 0:\ 2(5)\sin\theta - 8 = 0 \Rightarrow \theta = 53.13^\circ$$

6. Solve for the sag s:

$$\tan\theta = \frac{s}{4} \Rightarrow s = 4\tan 53.13^\circ = 5.33 \text{ ft} \qquad \textbf{Ans.}$$

Problem 3.9

A vertical force **P** is applied to the ends of the cord AB and spring AC. The spring has an unstretched length of 2-ft and the system is in equilibrium at angle θ. Draw the free-body diagram of the connecting knot at A and write down the equilibrium equations for the knot at A.

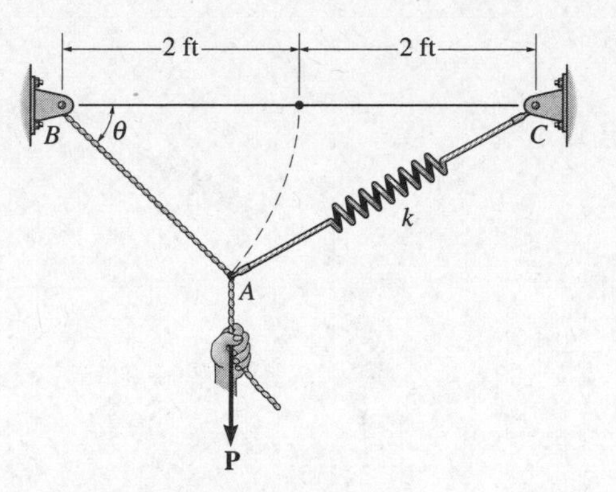

Solution

1. The knot at A has *negligible size* so that it can be modelled as a particle.
2. Imagine the knot at A to be separated or detached from the system.
3. The (detached) knot at A is subjected to three *external* forces. They are caused by:

   i. **ii.**

   iii.

4. Draw the free-body diagram of the (detached) knot showing all these forces labeled with their magnitudes and directions.

5. Establish an xy-axes system on the free-body diagram and write down the equilibrium equations in each of the x and y-directions

$+\uparrow \sum F_y = 0:$

$\underrightarrow{+} \sum F_x = 0:$

Problem 3.9

A vertical force **P** is applied to the ends of the cord AB and spring AC. The spring has an unstretched length of 2-ft and the system is in equilibrium at angle θ. Draw the free-body diagram of the connecting knot at A and write down the equilibrium equations for the knot at A.

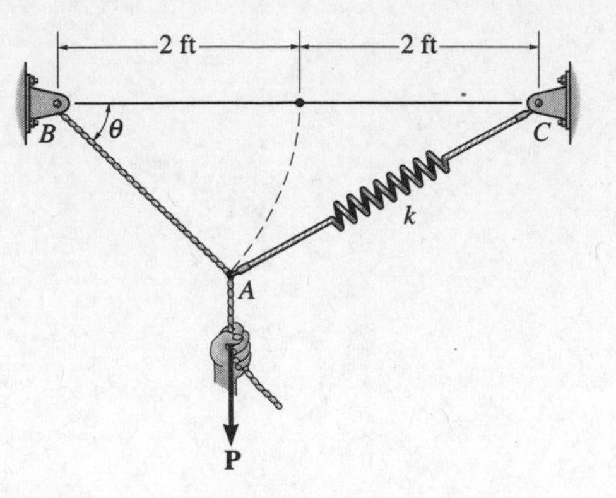

Solution

1. The knot at A has *negligible size* so that it can be modelled as a particle.
2. Imagine the knot at A to be separated or detached from the system.
3. The (detached) knot at A is subjected to three *external* forces. They are caused by:

   i. CORD AB **ii. SPRING** AC

   iii. Force P

4. Draw the free-body diagram of the (detached) knot showing all these forces labeled with their magnitudes and directions.

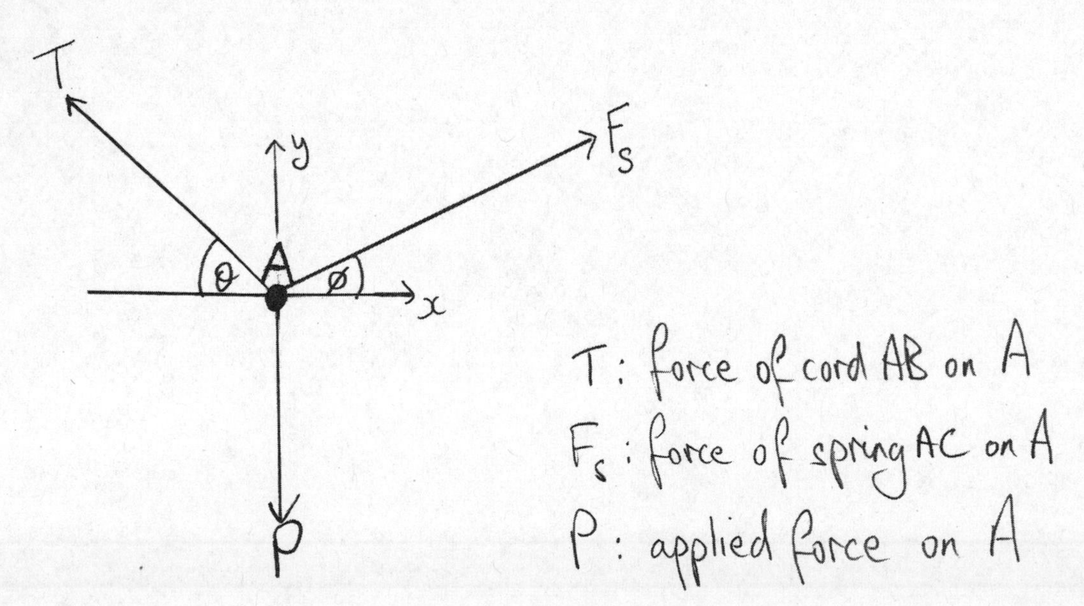

5. Establish an xy-axes system on the free-body diagram and write down the equilibrium equations in each of the x and y-directions

$$\overset{+}{\rightarrow} \sum F_x = 0: \; F_s \cos\phi - T\cos\theta = 0$$

$$+\uparrow \sum F_y = 0: \; T\sin\theta + F_s \sin\phi - P = 0$$

Ans.

Problem 3.10

The sling BAC is used to lift the 100-lb load with constant velocity. By drawing the free-body diagram for the ring at A, determine the magnitude of the force in the sling as a function of the angle θ.

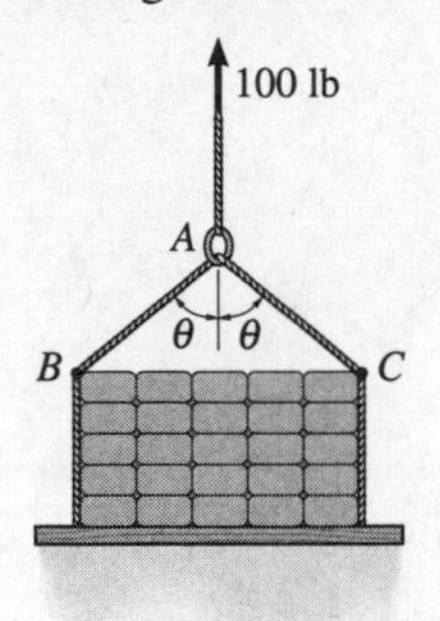

Solution

1. The ring at A has *negligible size* so that it can be modelled as a particle.
2. Imagine the ring at A to be separated or detached from the system.
3. The (detached) ring at A is subjected to three *external* forces. They are caused by:

   i. 　　　　**ii.**

   iii.

4. Draw the free-body diagram of the (detached) ring showing all these forces labeled with their magnitudes and directions. Include also any other information which may help when formulating the equilibrium equations for the ring.

5. Establish an xy-axes system on the free-body diagram and write down the equilibrium equations in the y-direction only (this is all that is required to solve this problem):

   $+\uparrow \sum F_y = 0$:

6. Solve for the magnitude of the force in the sling:

Problem 3.10

The sling BAC is used to lift the 100-lb load with constant velocity. By drawing the free-body diagram for the ring at A, determine the magnitude of the force in the sling as a function of the angle θ.

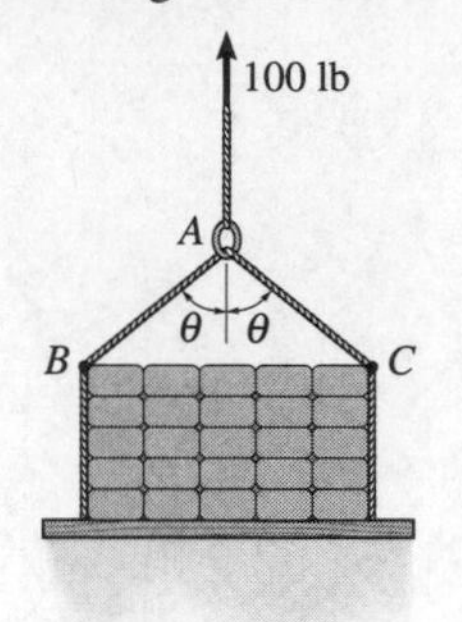

Solution

1. The ring at A has *negligible size* so that it can be modelled as a particle.
2. Imagine the ring at A to be separated or detached from the system.
3. The (detached) ring at A is subjected to three *external* forces. They are caused by:

   i. **CORD** AB

   ii. **CORD** AC

   iii. **CORD** AD *(weight of load)*

4. Draw the free-body diagram of the (detached) ring showing all these forces labeled with their magnitudes and directions. Include also any other information which may help when formulating the equilibrium equations for the ring.

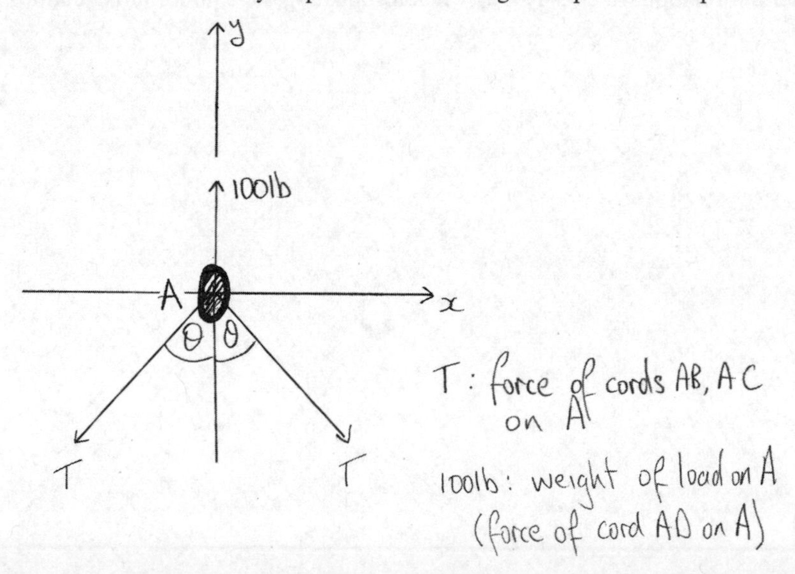

5. Establish an xy-axes system on the free-body diagram and write down the equilibrium equations in the y-direction only (this is all that is required to solve this problem):

   $+\uparrow \sum F_y = 0\text{: } 100 - 2T\cos\theta = 0$

6. Solve for the magnitude of the force in the sling:

$$T = \frac{50}{\cos\theta}$$ **Ans.**

Problem 3.11

When y is zero, the springs sustain a force of 60 lb. The applied vertical forces **F** and $-$**F** pull the point A away from B a distance of $y = 2$ ft. The cords CAD and CBD are attached to the rings at C and D. Draw the free-body diagrams for point A and ring C.

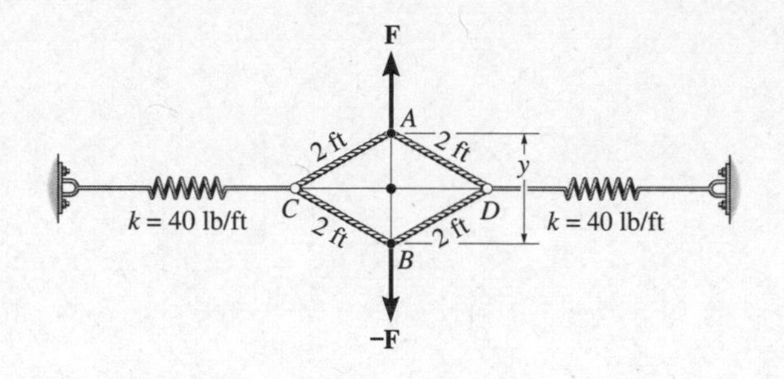

Solution

1. Imagine A and C to be separated or detached from the system.
2. Each of A and C is subjected to three *external* forces. For A, they are caused by:

   i. **ii.**

   iii.

   For C, they are caused by:

   i. **ii.**

   iii.

3. Draw the free-body diagrams of A and C showing all these forces labeled with their magnitudes and directions. You should also include any other available information e.g. lengths, angles etc. — which will help when formulating the equilibrium equations.

• A

o
C

Problem 3.11

When y is zero, the springs sustain a force of 60 lb. The applied vertical forces **F** and −**F** pull the point A away from B a distance of $y = 2$ ft. The cords CAD and CBD are attached to the rings at C and D. Draw the free-body diagrams for point A and ring C.

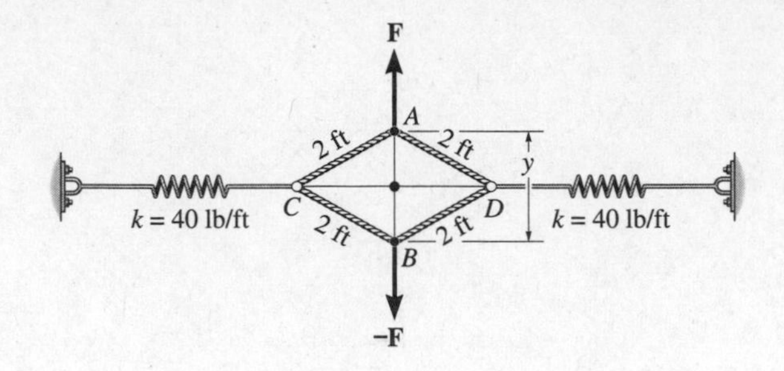

Solution

1. Imagine A and C to be separated or detached from the system.
2. Each of A and C is subjected to three *external* forces. For A, they are caused by:

   i. CORD AC **ii. CORD** AD

   iii. Force F

   For C, they are caused by:

   i. CORD AC **ii. CORD** CB

   iii. Spring attached at C
3. Draw the free-body diagrams of A and C showing all these forces labeled with their magnitudes and directions. You should also include any other available information e.g. lengths, angles etc. — which will help when formulating the equilibrium equations.

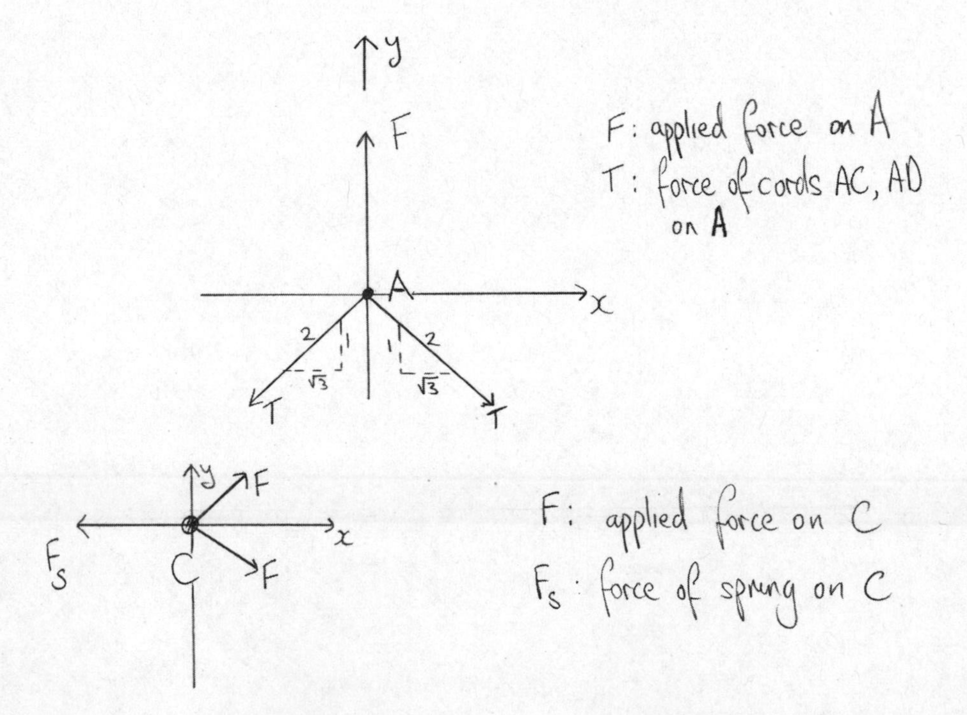

Problem 3.12

By drawing a free-body diagram for the ring at A, determine the maximum weight $\mathbf{W}$ that can be supported in the position shown if each cable AC and AB can support a maximum tension of 600 lb before it fails.

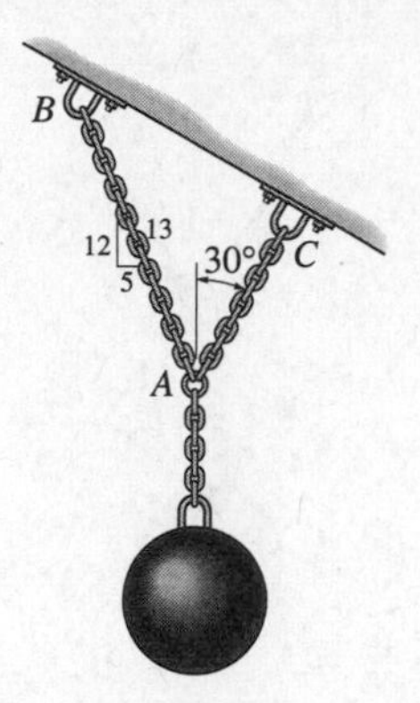

Solution

1. The ring at A has *negligible size* so that it can be modelled as a particle.
2. Imagine the ring at A to be separated or detached from the system.
3. The (detached) ring at A is subjected to three *external* forces. They are caused by:

   i. **ii.**

   iii.

4. Draw the free-body diagram of the (detached) ring showing all these forces labeled with their magnitudes and directions. Include any other relevant information e.g. lengths, angles etc.

AO

5. Establish an xy-axes system on the free-body diagram and write down the equilibrium equations in each of the x and y-directions

   $+\uparrow \sum F_y = 0:$

   $\overset{+}{\rightarrow} \sum F_x = 0:$

6. Set the tension in AB to the maximum of 600 lb and solve for the maximum weight $\mathbf{W}$:

Problem 3.12

By drawing a free-body diagram for the ring at A, determine the maximum weight $\mathbf{W}$ that can be supported in the position shown if each cable AC and AB can support a maximum tension of 600 lb before it fails.

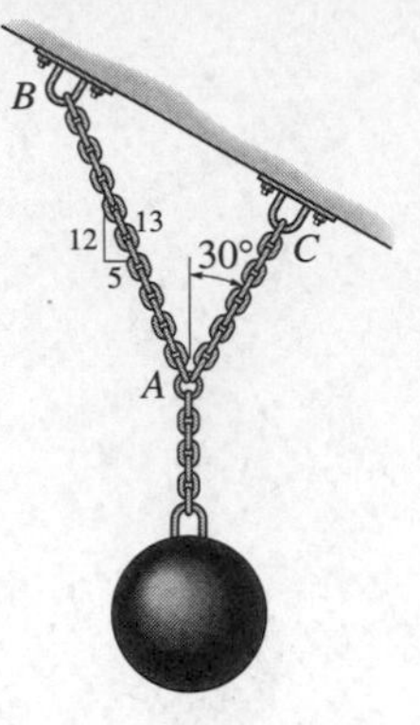

Solution

1. The ring at A has *negligible size* so that it can be modelled as a particle.
2. Imagine the ring at A to be separated or detached from the system.
3. The (detached) ring at A is subjected to three *external* forces. They are caused by:

   i. CABLE AB **ii. CABLE AB**

   iii. Weight of ball

4. Draw the free-body diagram of the (detached) ring showing all these forces labeled with their magnitudes and directions. Include any other relevant information e.g. lengths, angles etc.

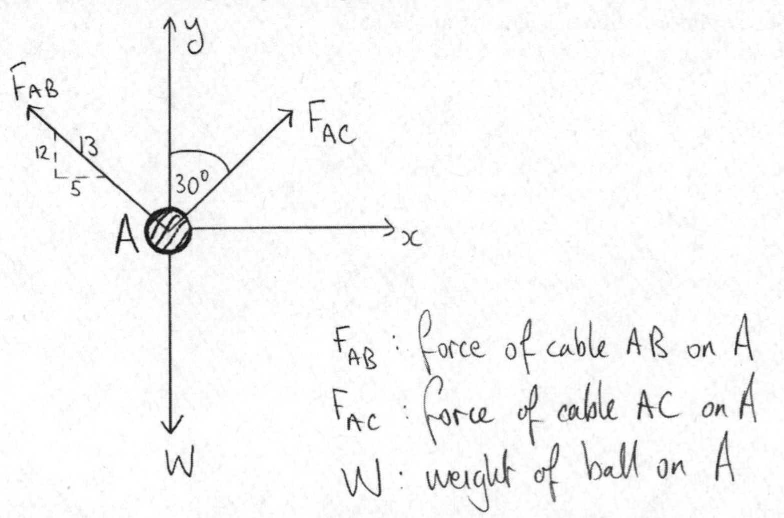

5. Establish an xy-axes system on the free-body diagram and write down the equilibrium equations in each of the x and y-directions

$$\overset{+}{\rightarrow} \sum F_x = 0: \quad -F_{AB}\left(\frac{5}{13}\right) + F_{AC}\sin 30^\circ = 0$$

$$+\uparrow \sum F_y = 0: \quad F_{AB}\left(\frac{12}{13}\right) + F_{AC}\cos 30^\circ - W = 0$$

6. Set F_{AB}, the tension in AB, to the maximum of 600 lb and solve for the maximum weight $\mathbf{W}$:

$$F_{AC} = 461.54 \text{ lb}(< 600 \text{ lb!!}), \quad \mathbf{W} = 953.55 \text{ lb} \downarrow$$ **Ans.**

Problem 3.13

The cords suspend the two *small* buckets in the equilibrium position shown. Draw the free-body diagrams for each of the points F and C.

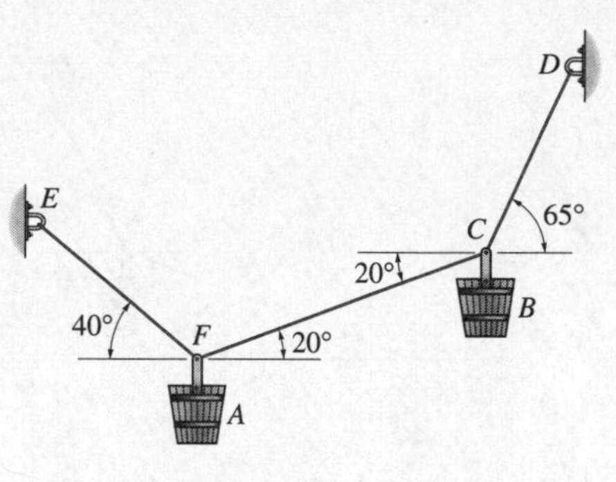

Solution

1. Imagine the points F and C to be separated or detached from the system.
2. Each of F and C is subjected to three *external* forces. For F, they are caused by:

   i. **ii.**

   iii.

   For C, they are caused by:

   i. **ii.**

   iii.

3. Draw the free-body diagrams of F and C showing all these forces labeled with their magnitudes and directions. You should also include any other available information e.g. lengths, angles etc. — which will help when formulating the equilibrium equations at these points.

F

C

Problem 3.13

The cords suspend the two *small* buckets in the equilibrium position shown. Draw the free-body diagrams for each of the points F and C.

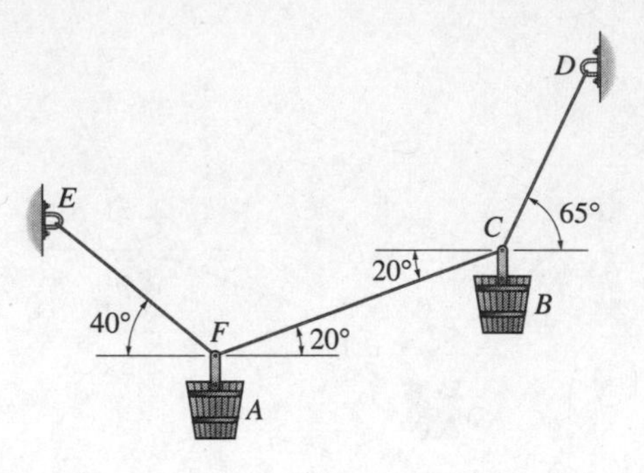

Solution

1. Imagine the points F and C to be separated or detached from the system.
2. Each of F and C is subjected to three *external* forces. For F, they are caused by:
   - **i. CABLE** FE
   - **ii. CABLE** FC
   - **iii. Weight of** A

   For C, they are caused by:
   - **i. CABLE** CF
   - **ii. CABLE** CD
   - **iii. Weight of** B
3. Draw the free-body diagrams of F and C showing all these forces labeled with their magnitudes and directions. You should also include any other available information e.g. lengths, angles etc. — which will help when formulating the equilibrium equations at these points.

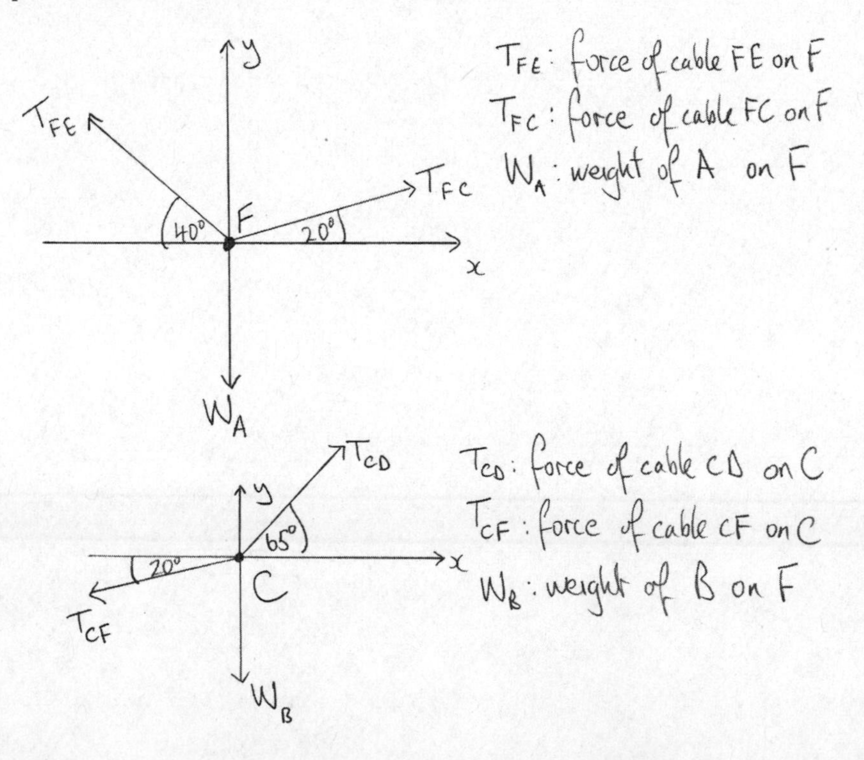

Problem 3.14

The 30-kg pipe is supported at A by a system of five cords. Draw the free-body diagrams for the rings at A and B when the system is in equilibrium.

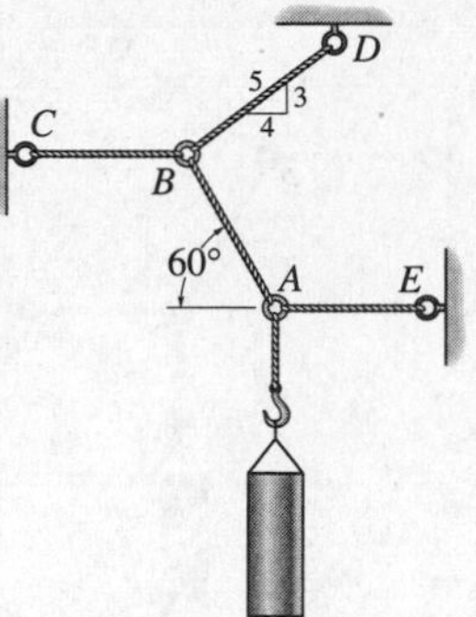

Solution

1. Imagine A and B to be separated or detached from the system.
2. Each of A and B is subjected to three *external* forces. For A, they are caused by:

   i. 　　　　**ii.**

   iii.

   For B, they are caused by:

   i. 　　　　**ii.**

   iii.

3. Draw the free-body diagrams of A and B showing all these forces labeled with their magnitudes and directions. You should also include any other available information e.g. lengths, angles etc. — which will help when formulating the equilibrium equations.

Problem 3.14

The 30-kg pipe is supported at *A* by a system of five cords. Draw the free-body diagrams for the rings at *A* and *B* when the system is in equilibrium.

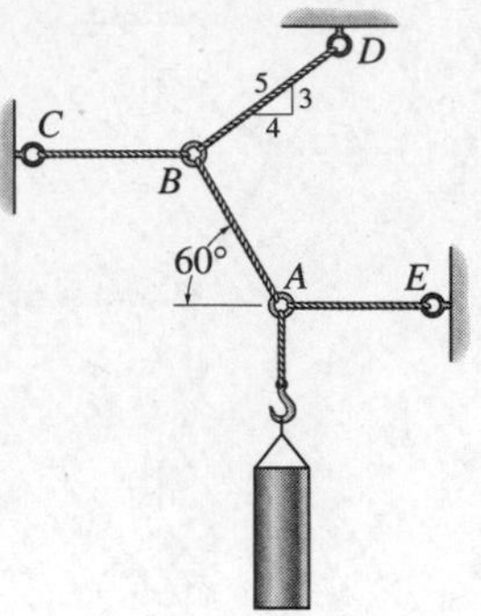

Solution

1. Imagine *A* and *B* to be separated or detached from the system.
2. Each of *A* and *B* is subjected to three *external* forces. For *A*, they are caused by:

   i. CABLE *AB* **ii. CABLE *AE***

   iii. Weight of Pipe

   For *B*, they are caused by:

   i. CABLE *BC* **ii. CABLE *BD***

   iii. CABLE *BA*
3. Draw the free-body diagrams of *A* and *B* showing all these forces labeled with their magnitudes and directions. You should also include any other available information e.g. lengths, angles etc. — which will help when formulating the equilibrium equations.

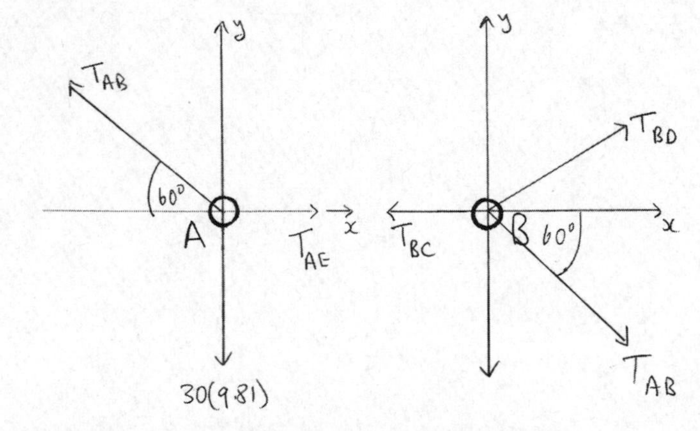

T_{AB}: force of cable AB (BA) on A(B)
T_{AE} : force of cable AE on A
T_{BD} : force of cable BD on B
T_{BC} : force of cable BC on B
30(9.81): weight of pipe on A

Problem 3.15

The cord AB has a length of 5 ft and is attached to the end B of the spring having a stiffness $k = 10$ lb/ft. The other end of the spring is attached to a roller C so that the spring remains horizontal as it stretches. If a 10-lb weight is suspended from B, use the free-body diagram for the ring at B to determine the necessary unstretched length of the spring, so that $\theta = 40°$ for equilibrium.

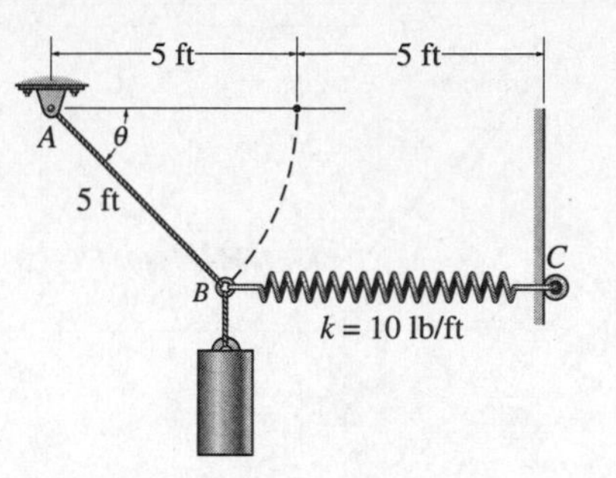

Solution

1. Imagine the ring at B to be separated or detached from the system.
2. The (detached) ring at B is subjected to three *external* forces caused by:

   i. **ii.**

   iii.

3. Draw the free-body diagram of the (detached) ring showing all these forces labeled with their magnitudes and directions. Include any other relevant information e.g. lengths, angles etc.

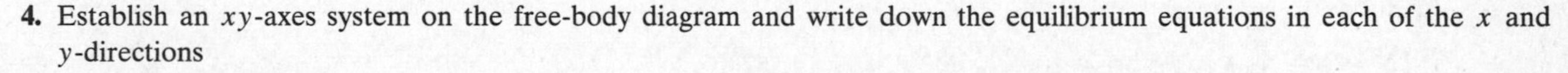

4. Establish an xy-axes system on the free-body diagram and write down the equilibrium equations in each of the x and y-directions

   $+\uparrow \sum F_y = 0:$

   $\underset{\rightarrow}{+} \sum F_x = 0:$

5. Determine the stretch in the spring BC and solve for the necessary unstretched length:

Problem 3.15

The cord AB has a length of 5 ft and is attached to the end B of the spring having a stiffness $k = 10$ lb/ft. The other end of the spring is attached to a roller C so that the spring remains horizontal as it stretches. If a 10-lb weight is suspended from B, use the free-body diagram for the ring at B to determine the necessary unstretched length of the spring, so that $\theta = 40°$ for equilibrium.

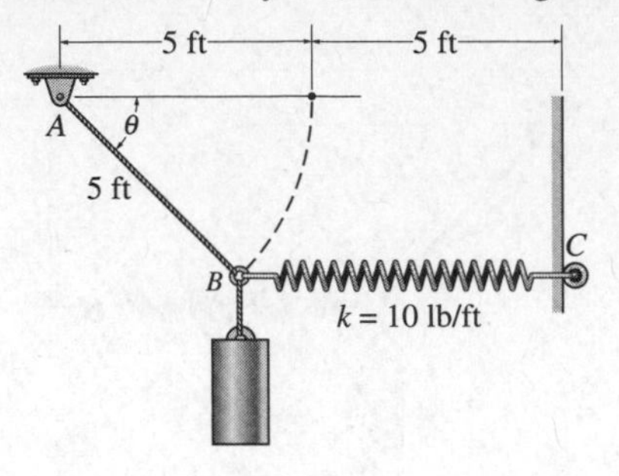

Solution

1. Imagine the ring at B to be separated or detached from the system.
2. The (detached) ring at B is subjected to three *external* forces caused by:

   i. CABLE AB **ii. SPRING BC**

   iii. 10 lb Weight

3. Draw the free-body diagram of the (detached) ring showing all these forces labeled with their magnitudes and directions. Include any other relevant information e.g. lengths, angles etc.

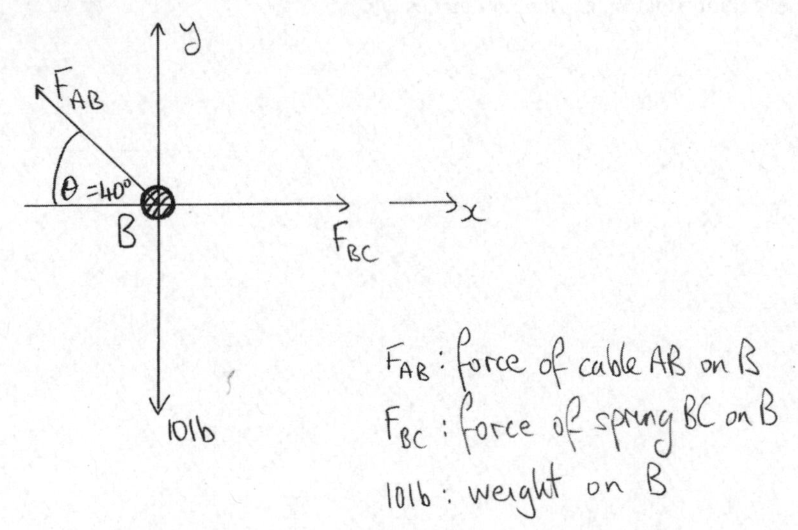

4. Establish an xy-axes system on the free-body diagram and write down the equilibrium equations in each of the x and y-directions

$$+\uparrow \sum F_y = 0: \; F_{AB}\sin 40° - 10 = 0 \Rightarrow F_{AB} = 15.557 \text{ lb}$$

$$\overset{+}{\rightarrow} \sum F_x = 0: \; F_{BC} - F_{AB}\cos 40° = 0 \Rightarrow F_{BC} = 11.918 \text{ lb}$$

5. Determine the stretch in the spring BC and solve for the necessary unstretched length l:

$$F_{BC} = kx \Rightarrow x = \frac{11.918}{10} = 1.1918 \text{ ft (stretch in } BC\text{)}$$

$$BC = 5 + (5 - 5\cos 40°) = 6.17 \text{ ft}$$

$$l = 6.17 - 1.1918 = 4.98 \text{ ft}$$

Ans.

3.2 Free-Body Diagrams in the Equilibrium of a Rigid Body

Problem 3.16

Draw the free-body diagram of the 50-kg uniform pipe, which is supported by the smooth contacts at A and B.

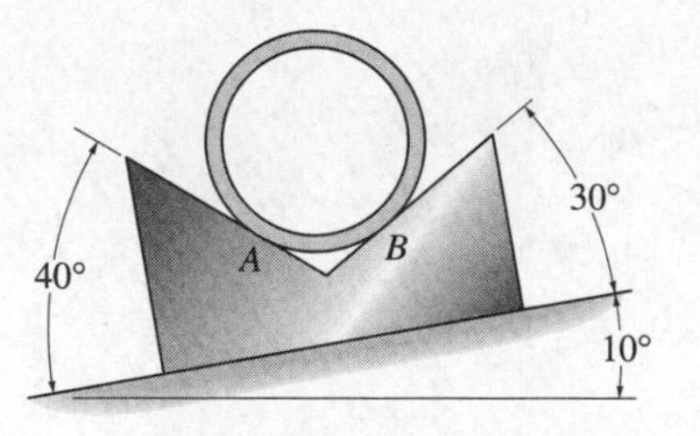

Solution

1. Imagine the pipe to be separated or detached from the system.
2. The supports at A and B are smooth contacts. Use Table 2.1 (6) to determine the number and types of reactions *acting on the pipe* at A and B.
3. The pipe is subjected to three *external* forces (don't forget the weight!). They are caused by:

   i. **ii.**
   iii.
4. Draw the free-body diagram of the (detached) pipe showing all these forces labeled with their magnitudes and directions. *Assume* the sense of the vectors representing the *reactions acting on the pipe* (the correct sense will always emerge from the equilibrium equations for the pipe). Include any other relevant information e.g. lengths, angles etc. which may help when formulating the equilibrium equations (including the moment equation) for the pipe.

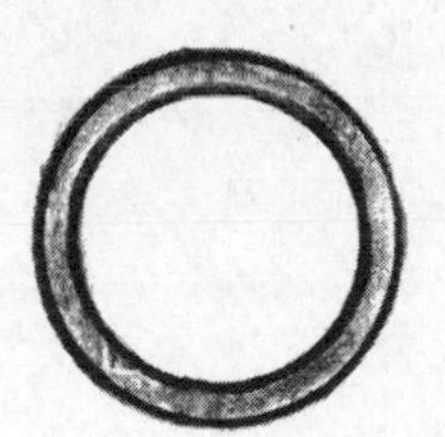

Problem 3.16

Draw the free-body diagram of the 50-kg uniform pipe, which is supported by the smooth contacts at A and B.

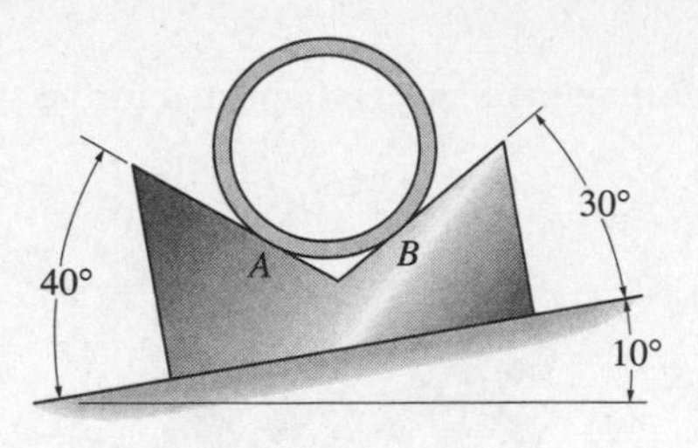

Solution

1. Imagine the pipe to be separated or detached from the system.
2. The supports at A and B are smooth contacts. Use Table 2.1 (6) to determine the number and types of reactions *acting on the pipe* at A and B.
3. The pipe is subjected to three *external* forces (don't forget the weight!). They are caused by:

   i. The reaction at A

   ii. The reaction at B

   iii. The weight of the pipe

4. Draw the free-body diagram of the (detached) pipe showing all these forces labeled with their magnitudes and directions. *Assume* the sense of the vectors representing the *reactions acting on the pipe* (the correct sense will always emerge from the equilibrium equations for the pipe). Include any other relevant information e.g. lengths, angles etc. which may help when formulating the equilibrium equations (including the moment equation) for the pipe.

50(9.81)N

N_A, N_B: force of contacting surface on pipe

50(9.81): gravity (weight) on pipe

Problem 3.17

Draw the free-body diagram of the hand punch, which is pinned at A and bears down on the smooth surface at B. Neglect the weight of the punch.

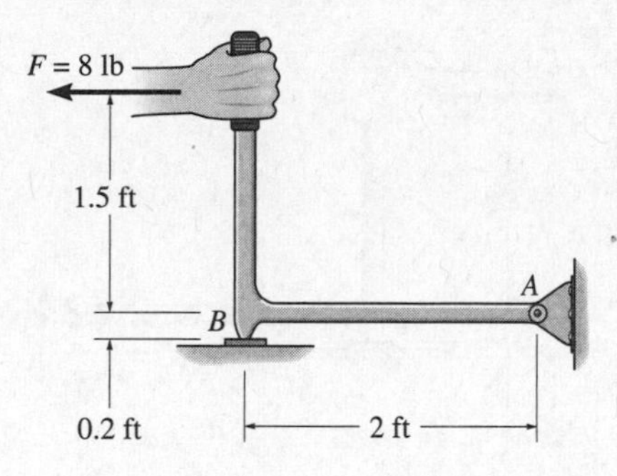

Solution

1. Imagine the hand punch to be separated or detached from the system.
2. The support at B is a smooth contact. The punch is (smoothly) pin-connected at A. Use Table 2.1 (6) and (8) to determine the number and types of reactions *acting on the pipe* at A and B.
3. The punch is subjected to four *external* forces. They are caused by:

   i. **ii.**

   iii. **iv.**

4. Draw the free-body diagram of the (detached) punch showing all these forces labeled with their magnitudes and directions. *Assume* the sense of the vectors representing the *reactions acting on the punch* (the correct sense will always emerge from the equilibrium equations for the punch). Include any other relevant information e.g. lengths, angles etc. which may help when formulating the equilibrium equations (including the moment equation) for the punch.

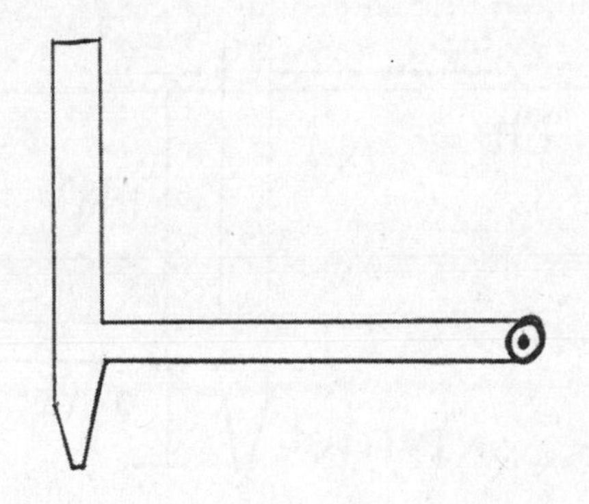

Problem 3.17

Draw the free-body diagram of the hand punch, which is pinned at A and bears down on the smooth surface at B. Neglect the weight of the punch.

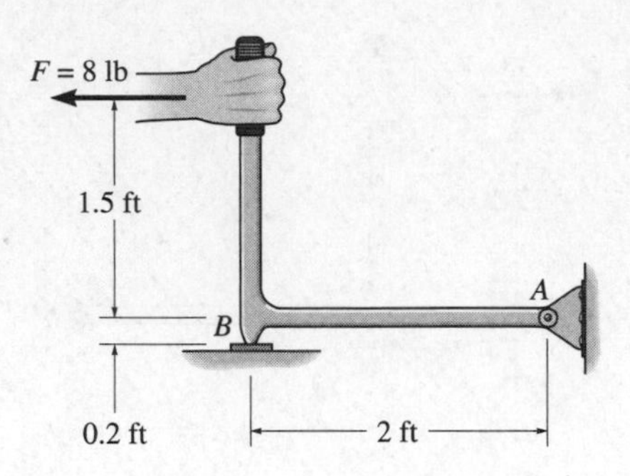

Solution

1. Imagine the hand punch to be separated or detached from the system.
2. The support at B is a smooth contact. The punch is (smoothly) pin-connected at A. Use Table 2.1 (6) and (8) to determine the number and types of reactions *acting on the pipe* at A and B.
3. The punch is subjected to four *external* forces. They are caused by:

   i. The force F

   ii. The reaction at B

   iii. & iv. The two reactions at A

4. Draw the free-body diagram of the (detached) punch showing all these forces labeled with their magnitudes and directions. *Assume* the sense of the vectors representing the *reactions acting on the punch* (the correct sense will always emerge from the equilibrium equations for the punch). Include any other relevant information e.g. lengths, angles etc. which may help when formulating the equilibrium equations (including the moment equation) for the punch.

$F = 8\text{lb}$

1.5ft

0.2ft

A_x

A_y

B_y

2ft

B_y: force of surface on punch

A_x, A_y: force of pin on member

F: applied force

Problem 3.18

Draw the free-body diagram of the jib crane AB, which is pin-connected at A and supported by member (link) BC. Neglect the weight of the crane.

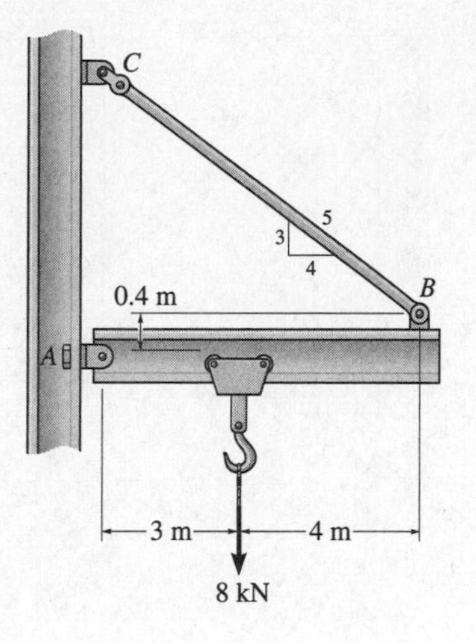

Solution

1. Imagine the jib crane AB to be separated or detached from the system.
2. There is a link support at B and the jib crane is (smoothly) pinned at A. Use Table 2.1 (2) and (8) to determine the number and types of reactions *acting on the jib crane* at A and B.
3. The jib crane is subjected to four *external* forces. They are caused by:

   i. 　　　　　　　　**ii.**

   iii. 　　　　　　　**iv.**

4. Draw the free-body diagram of the (detached) crane showing all these forces labeled with their magnitudes and directions. *Assume* the sense of the vectors representing the *reactions acting on the crane* (the correct sense will always emerge from the equilibrium equations for the crane). Include any other relevant information e.g. lengths, angles etc. which may help when formulating the equilibrium equations (including the moment equation) for the jib crane.

Problem 3.18

Draw the free-body diagram of the jib crane AB, which is pin-connected at A and supported by member (link) BC. Neglect the weight of the crane.

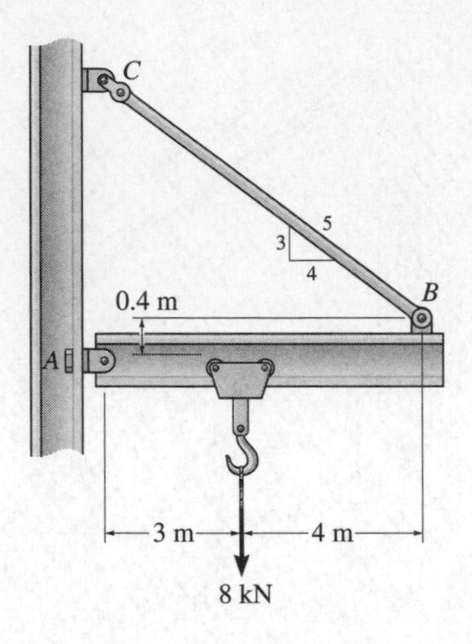

Solution

1. Imagine the jib crane AB to be separated or detached from the system.
2. There is a link support at B and the jib crane is (smoothly) pinned at A. Use Table 2.1 (2) and (8) to determine the number and types of reactions *acting on the jib crane* at A and B.
3. The jib crane is subjected to four *external* forces. They are caused by:

   i. & ii. The reactions at A **iii. The reaction at** B

   iv. The 8 kN load

4. Draw the free-body diagram of the (detached) crane showing all these forces labeled with their magnitudes and directions. *Assume* the sense of the vectors representing the *reactions acting on the crane* (the correct sense will always emerge from the equilibrium equations for the crane). Include any other relevant information e.g. lengths, angles etc. which may help when formulating the equilibrium equations (including the moment equation) for the jib crane.

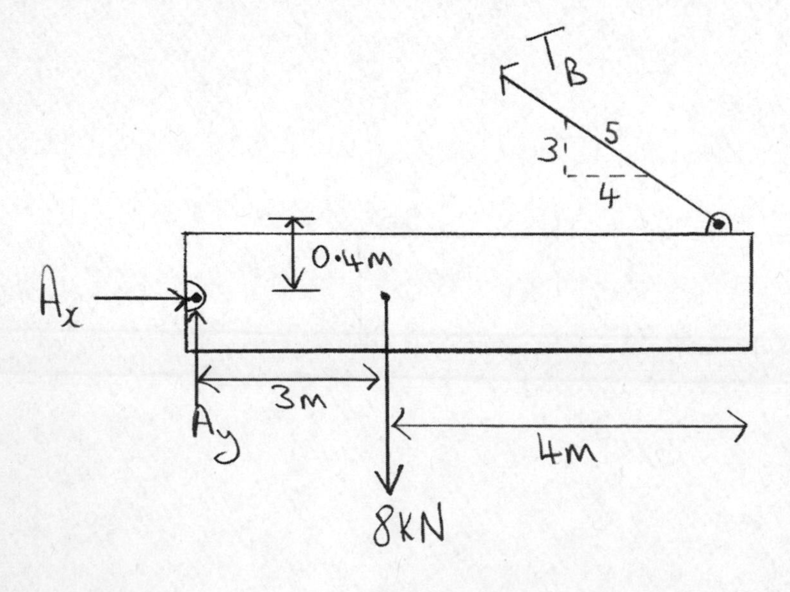

A_x, A_y: force of pin on AB
8kN : applied load
T_B : force of link BC on AB

Problem 3.19

Draw the free-body diagram of the dumpster D of the truck, which has a weight of 5000 lb and a center of gravity at G. It is supported by a pin at A and a pin-connected hydraulic cylinder BC (short link).

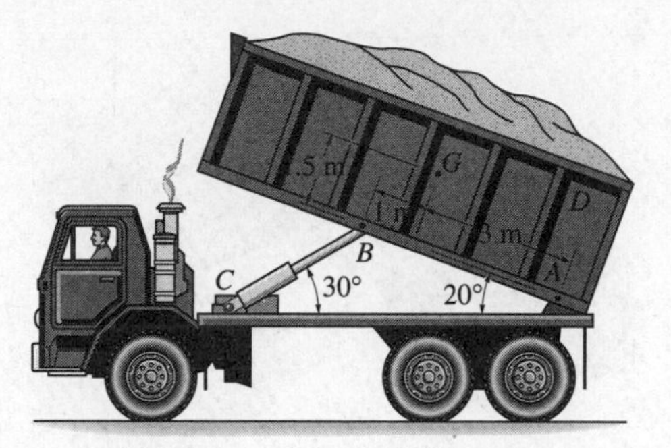

Solution

1. Imagine the dumpster D to be separated or detached from the truck.
2. There is a pin support at A and the dumpster is supported by a short link support at B. Use Table 2.1 (2) and (8) to determine the number and types of reactions *acting on the dumpster* at A and B.
3. The dumpster is subjected to four *external* forces. They are caused by:

   i. **ii.**

   iii. **iv.**

4. Draw the free-body diagram of the (detached) dumpster showing all these forces labeled with their magnitudes and directions. *Assume* the sense of the vectors representing the *reactions acting on the dumpster* (the correct sense will always emerge from the equilibrium equations for the dumpster). Include any other relevant information e.g. lengths, angles etc. which may help when formulating the equilibrium equations (including the moment equation) for the dumpster.

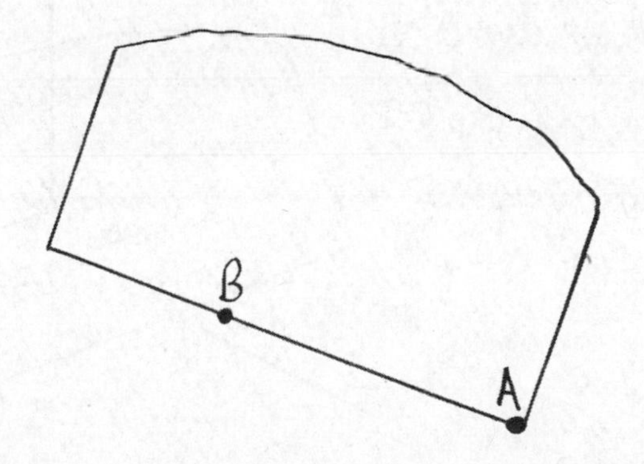

Problem 3.19

Draw the free-body diagram of the dumpster D of the truck, which has a weight of 5000 lb and a center of gravity at G. It is supported by a pin at A and a pin-connected hydraulic cylinder BC (short link).

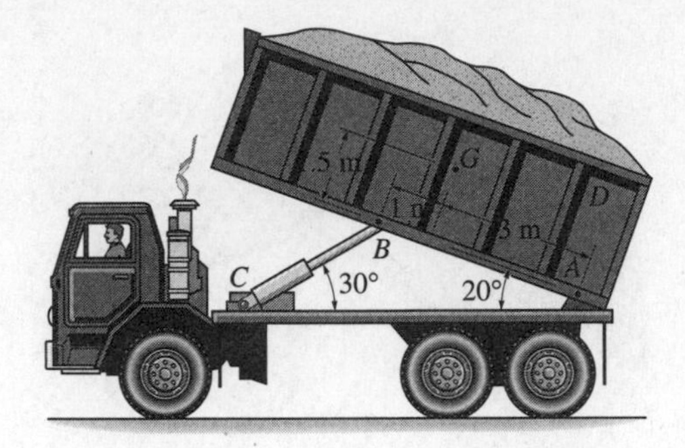

Solution

1. Imagine the dumpster D to be separated or detached from the truck.
2. There is a pin support at A and the dumpster is supported by a short link support at B. Use Table 2.1 (2) and (8) to determine the number and types of reactions *acting on the dumpster* at A and B.
3. The dumpster is subjected to four *external* forces. They are caused by:

   i. & ii. The reactions at A

   iii. The reaction at B

   iv. The weight of the dumpster
4. Draw the free-body diagram of the (detached) dumpster showing all these forces labeled with their magnitudes and directions. *Assume* the sense of the vectors representing the *reactions acting on the dumpster* (the correct sense will always emerge from the equilibrium equations for the dumpster). Include any other relevant information e.g. lengths, angles etc. which may help when formulating the equilibrium equations (including the moment equation) for the dumpster.

T_{CB}: force of link on dumpster
A_x, A_y: force of pin on dumpster
5000lb: weight on dumpster

Problem 3.20

Draw the free-body diagram of the link CAB, which is pin-connected at A and rests on the smooth cam at B. Neglect the weight of the link.

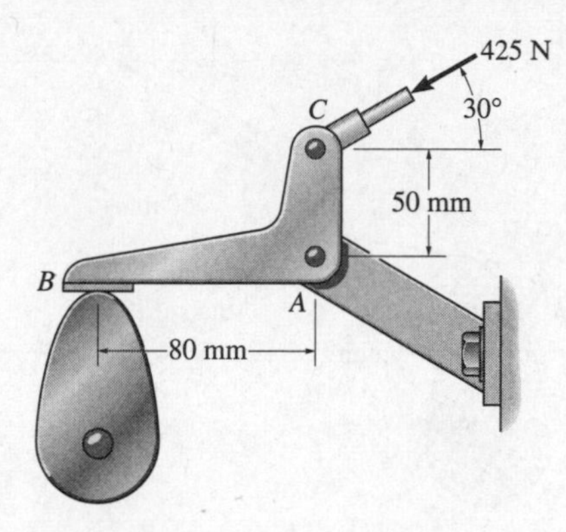

Solution

1. Imagine the link CAB to be separated or detached from the mechanism.
2. There is a pin connection at A and the link rests on the smooth surface at B. Use Table 2.1 (6) and (8) to determine the number and types of reactions *acting on the link* at A and B.
3. The link is subjected to four *external* forces. They are caused by:

   i. **ii.**

   iii. **iv.**

4. Draw the free-body diagram of the (detached) link showing all these forces labeled with their magnitudes and directions. *Assume* the sense of the vectors representing the *reactions acting on the link* (the correct sense will always emerge from the equilibrium equations for the link). Include any other relevant information e.g. lengths, angles etc. which may help when formulating the equilibrium equations (including the moment equation) for the link.

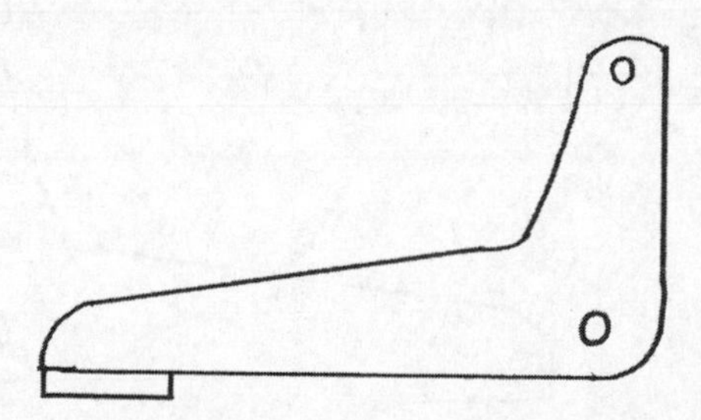

Problem 3.20

Draw the free-body diagram of the link CAB, which is pin-connected at A and rests on the smooth cam at B. Neglect the weight of the link.

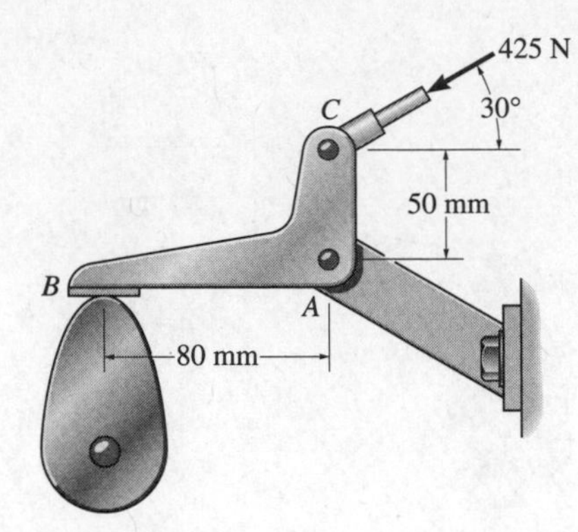

Solution

1. Imagine the link CAB to be separated or detached from the mechanism.
2. There is a pin connection at A and the link rests on the smooth surface at B. Use Table 2.1 (6) and (8) to determine the number and types of reactions *acting on the link* at A and B.
3. The link is subjected to four *external* forces. They are caused by:

   i. & ii. The reactions at A **iii. The reaction at** B

   iv. The 425 N load at C

4. Draw the free-body diagram of the (detached) link showing all these forces labeled with their magnitudes and directions. *Assume* the sense of the vectors representing the *reactions acting on the link* (the correct sense will always emerge from the equilibrium equations for the link). Include any other relevant information e.g. lengths, angles etc. which may help when formulating the equilibrium equations (including the moment equation) for the link.

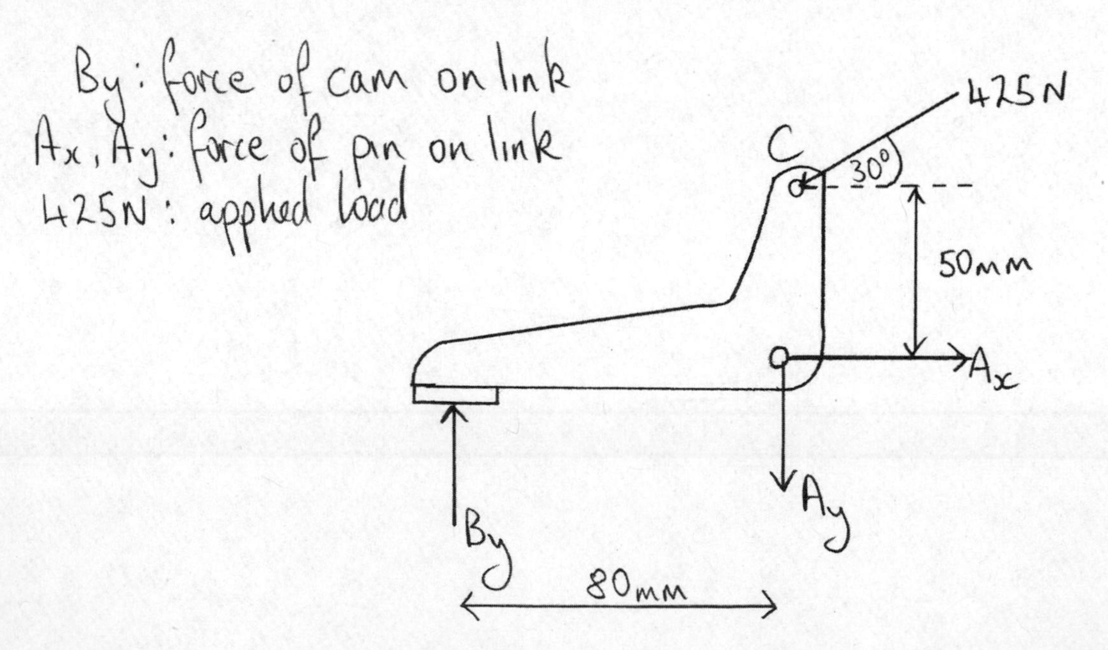

Problem 3.21

Draw the free-body diagram of the uniform pipe which has a mass of 100 kg and a center of mass at G. The supports A, B and C are smooth.

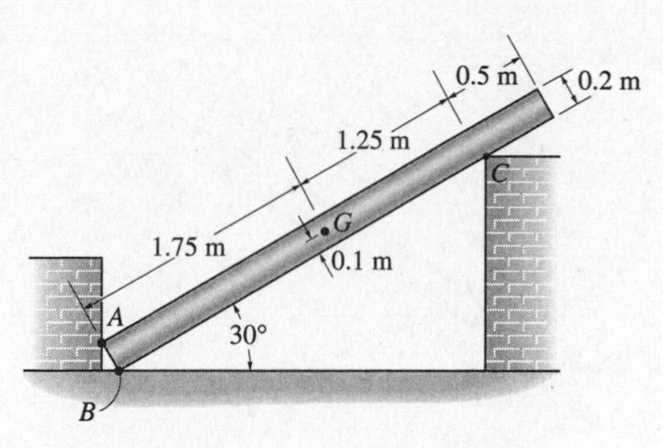

Solution

1. Imagine the pipe to be separated or detached from the system.
2. The pipe rests on smooth surfaces at A, B and C. Use Table 2.1 (6) to determine the number and types of reactions *acting on the pipe* at A, B and C.
3. The pipe is subjected to four *external* forces. They are caused by:

   i. **ii.**

   iii. **iv.**

4. Draw the free-body diagram of the (detached) pipe showing all these forces labeled with their magnitudes and directions. *Assume* the sense of the vectors representing the *reactions acting on the pipe* (the correct sense will always emerge from the equilibrium equations for the pipe). Include any other relevant information e.g. lengths, angles etc. which may help when formulating the equilibrium equations for the pipe.

Problem 3.21.

Draw the free-body diagram of the uniform pipe which has a mass of 100 kg and a center of mass at G. The supports A, B and C are smooth.

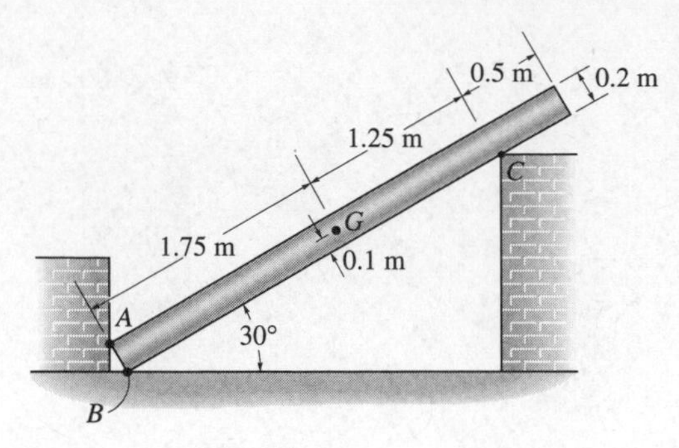

Solution

1. Imagine the pipe to be separated or detached from the system.
2. The pipe rests on smooth surfaces at A, B and C. Use Table 2.1 (6) to determine the number and types of reactions *acting on the pipe* at A, B and C.
3. The pipe is subjected to four *external* forces. They are caused by:

   i. The reaction at A

   ii. The reaction at B

   iii. The reaction at C

   iv. Pipe's weight
4. Draw the free-body diagram of the (detached) pipe showing all these forces labeled with their magnitudes and directions. *Assume* the sense of the vectors representing the *reactions acting on the pipe* (the correct sense will always emerge from the equilibrium equations for the pipe). Include any other relevant information e.g. lengths, angles etc. which may help when formulating the equilibrium equations for the pipe.

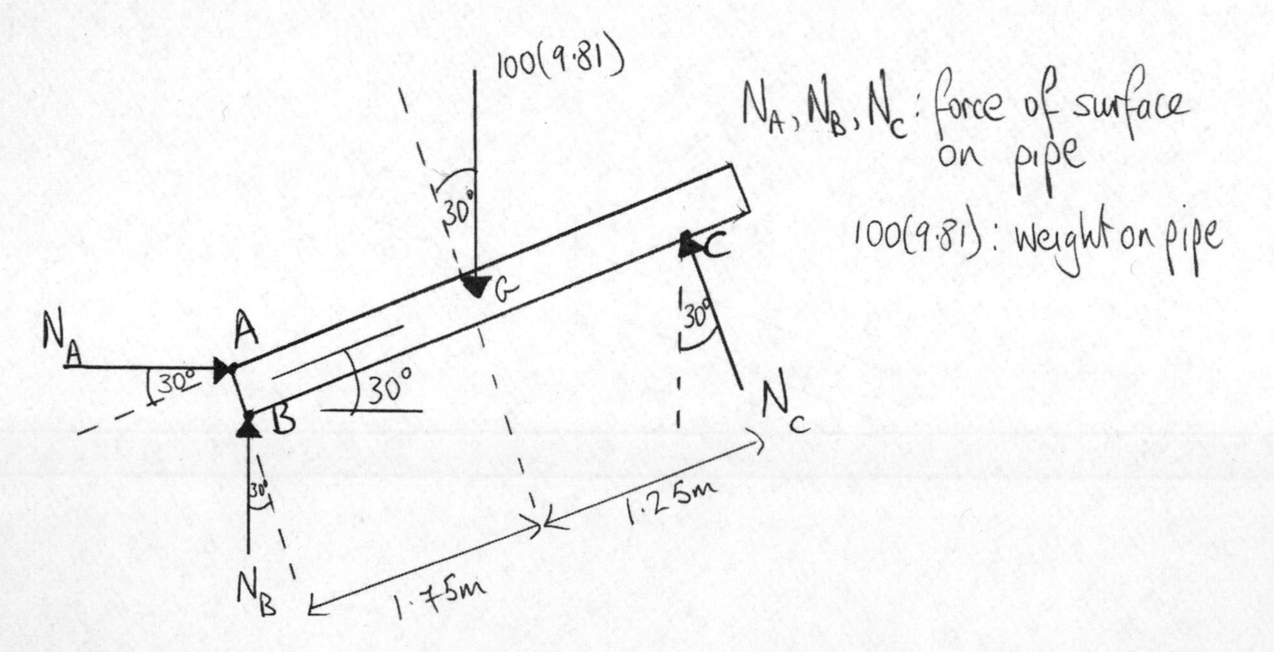

Problem 3.22

Draw the free-body diagram of the beam, which is pin-supported at A and rests on the smooth incline at B. Neglect the weight of the beam.

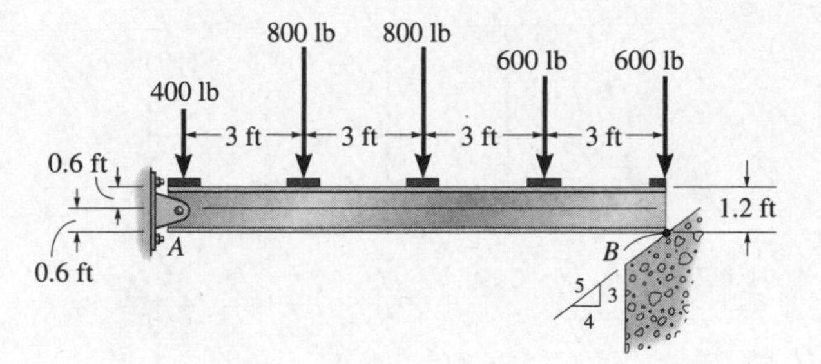

Solution

1. Imagine the beam to be separated or detached from the system.
2. There is a pin connection at A and the beam rests on the smooth (inclined) surface at B. Use Table 2.1 (6) and (8) to determine the number and types of reactions *acting on the beam* at A and B.
3. In addition to the forces shown in the figure, the beam is subjected to three *external* forces. They are caused by:

   i. **ii.**

   iii.

4. Draw the free-body diagram of the (detached) beam showing all these forces labeled with their magnitudes and directions. *Assume* the sense of the vectors representing the *reactions acting on the beam* (the correct sense will always emerge from the equilibrium equations for the beam). Include any other relevant information e.g. lengths, angles etc. which may help when formulating the equilibrium equations (including the moment equation) for the beam.

Problem 3.22

Draw the free-body diagram of the beam, which is pin-supported at A and rests on the smooth incline at B. Neglect the weight of the beam.

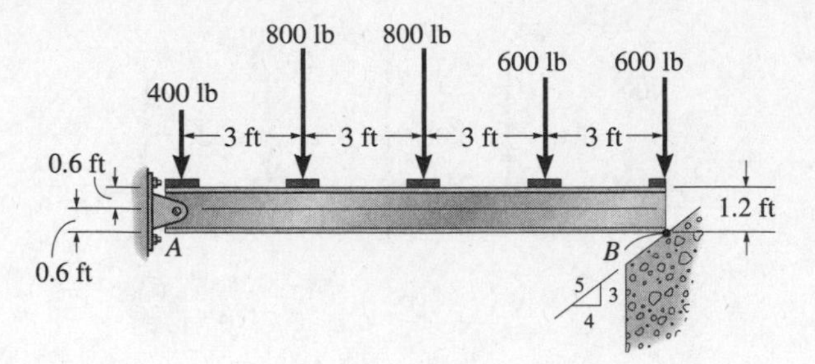

Solution

1. Imagine the beam to be separated or detached from the system.
2. There is a pin connection at A and the beam rests on the smooth (inclined) surface at B. Use Table 2.1 (6) and (8) to determine the number and types of reactions *acting on the beam* at A and B.
3. In addition to the forces shown in the figure, the beam is subjected to three *external* forces. They are caused by:

   i. & ii. The reactions at A **iii. The reaction at** B

4. Draw the free-body diagram of the (detached) beam showing all these forces labeled with their magnitudes and directions. *Assume* the sense of the vectors representing the *reactions acting on the beam* (the correct sense will always emerge from the equilibrium equations for the beam). Include any other relevant information e.g. lengths, angles etc. which may help when formulating the equilibrium equations (including the moment equation) for the beam.

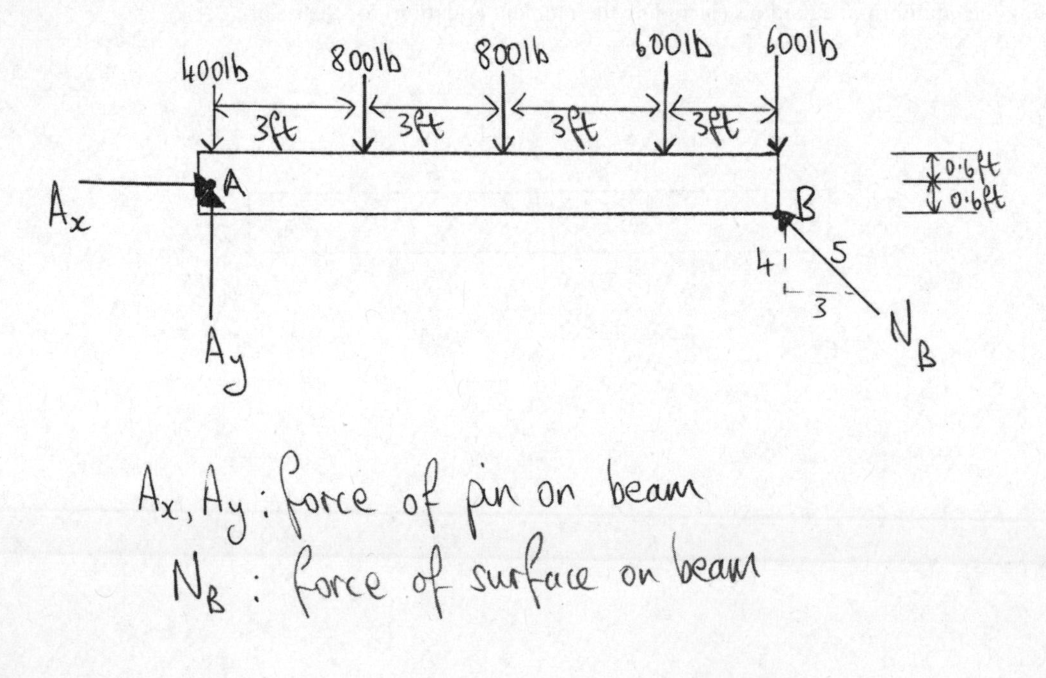

Problem 3.23

Draw the free-body diagram of the member ABC, which is supported by a pin at A and a horizontal short link BD. Neglect the weight of ABC.

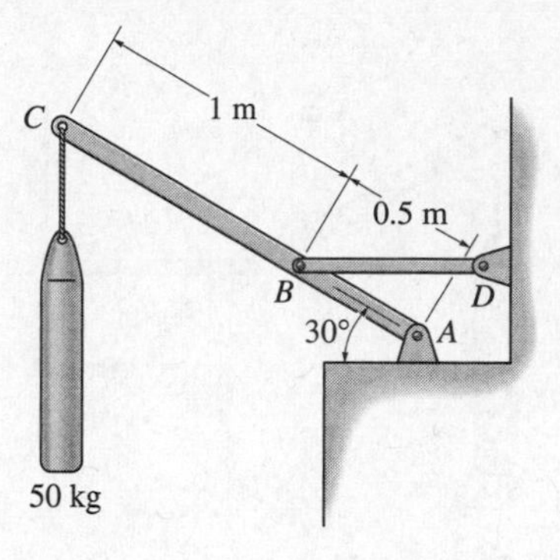

Solution

1. Imagine the member ABC to be separated or detached from the system.
2. There is a pin support at A and the member is supported by a horizontal short link at B. Use Table 2.1 (2) and (8) to determine the number and types of reactions *acting on the member* at A and B.
3. The member is subjected to four *external* forces. They are caused by:

   i. **ii.**

   iii. **iv.**
4. Draw the free-body diagram of the (detached) member showing all these forces labeled with their magnitudes and directions. *Assume* the sense of the vectors representing the *reactions acting on the member* (the correct sense will always emerge from the equilibrium equations for the member). Include any other relevant information e.g. lengths, angles etc. which may help when formulating the equilibrium equations (including the moment equation) for the member.

Problem 3.23

Draw the free-body diagram of the member ABC, which is supported by a pin at A and a horizontal short link BD. Neglect the weight of ABC.

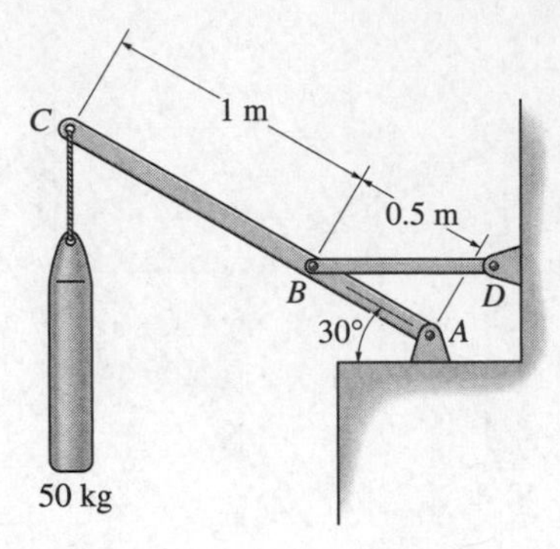

Solution

1. Imagine the member ABC to be separated or detached from the system.
2. There is a pin support at A and the member is supported by a horizontal short link at B. Use Table 2.1 (2) and (8) to determine the number and types of reactions *acting on the member* at A and B.
3. The member is subjected to four *external* forces. They are caused by:

   i. & ii. The reactions at A

   iii. The reaction at B

   iv. The weight at C
4. Draw the free-body diagram of the (detached) member showing all these forces labeled with their magnitudes and directions. *Assume* the sense of the vectors representing the *reactions acting on the member* (the correct sense will always emerge from the equilibrium equations for the member). Include any other relevant information e.g. lengths, angles etc. which may help when formulating the equilibrium equations (including the moment equation) for the member.

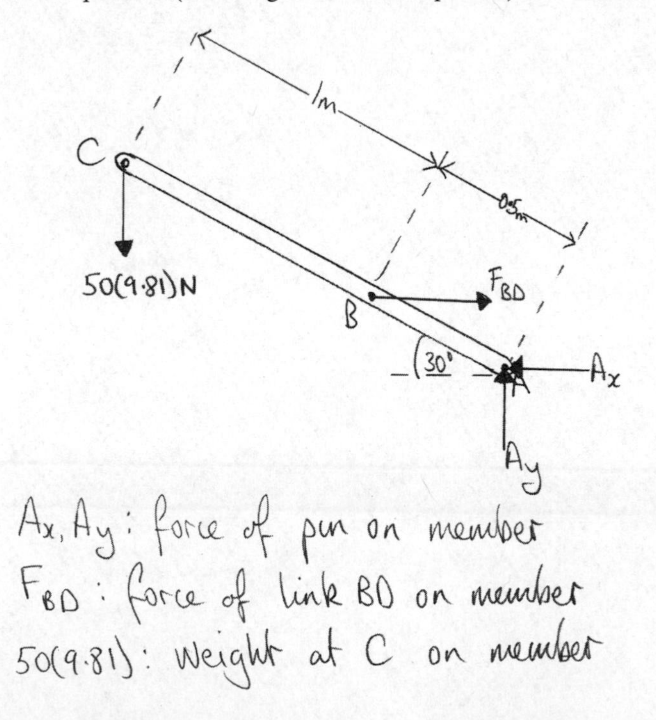

Problem 3.24

Draw the free-body diagram of the beam. The support at B is smooth. Neglect the weight of the beam.

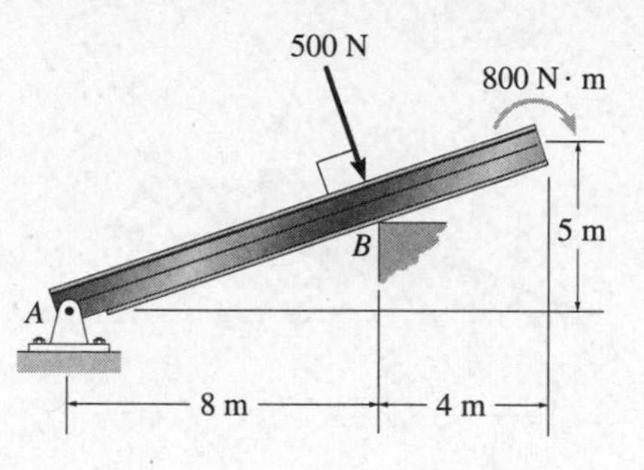

Solution

1. Imagine the beam to be separated or detached from the system.
2. There is a pin support at A and a smooth contact support at B. Use Table 2.1 (6) and (8) to determine the number and types of reactions *acting on the member* at A and B.
3. In addition to those shown in the figure, the member is subjected to three *external* forces. They are caused by:

   i. **ii.**

   iii.

4. Draw the free-body diagram of the (detached) member showing all these forces and any external applied couple moments labeled with their magnitudes and directions. *Assume* the sense of the vectors representing the *reactions acting on the member* (the correct sense will always emerge from the equilibrium equations for the member). Include any other relevant information e.g. lengths, angles etc. which may help when formulating the equilibrium equations (including the moment equation) for the member.

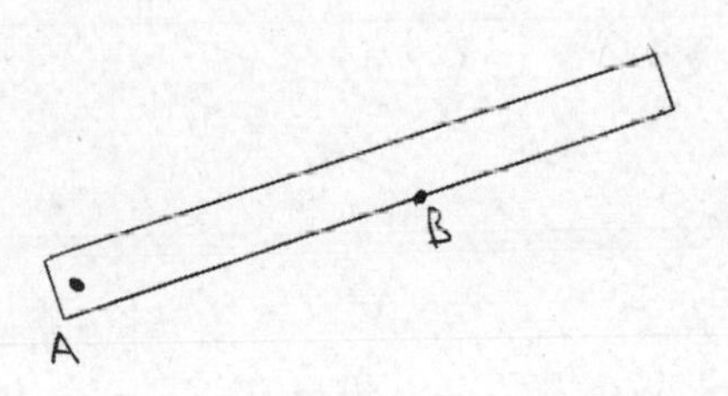

Problem 3.24

Draw the free-body diagram of the beam. The support at B is smooth. Neglect the weight of the beam.

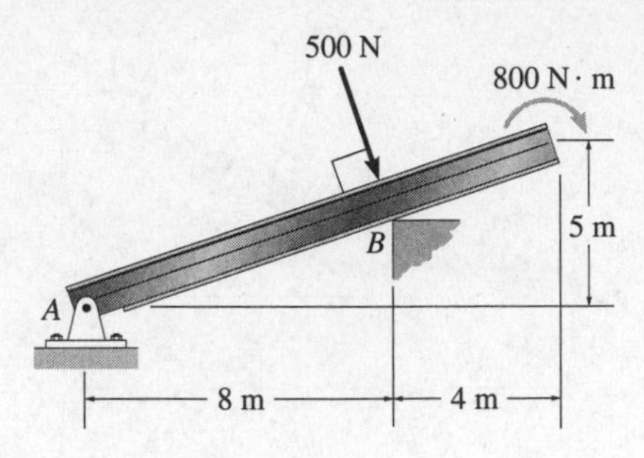

Solution

1. Imagine the beam to be separated or detached from the system.
2. There is a pin support at A and a smooth contact support at B. Use Table 2.1 (6) and (8) to determine the number and types of reactions *acting on the member* at A and B.
3. In addition to those shown in the figure, the member is subjected to three *external* forces. They are caused by:

   i. & ii. The reactions at A **iii. The reaction at B**

4. Draw the free-body diagram of the (detached) member showing all these forces and any external applied couple moments labeled with their magnitudes and directions. *Assume* the sense of the vectors representing the *reactions acting on the member* (the correct sense will always emerge from the equilibrium equations for the member). Include any other relevant information e.g. lengths, angles etc. which may help when formulating the equilibrium equations (including the moment equation) for the member.

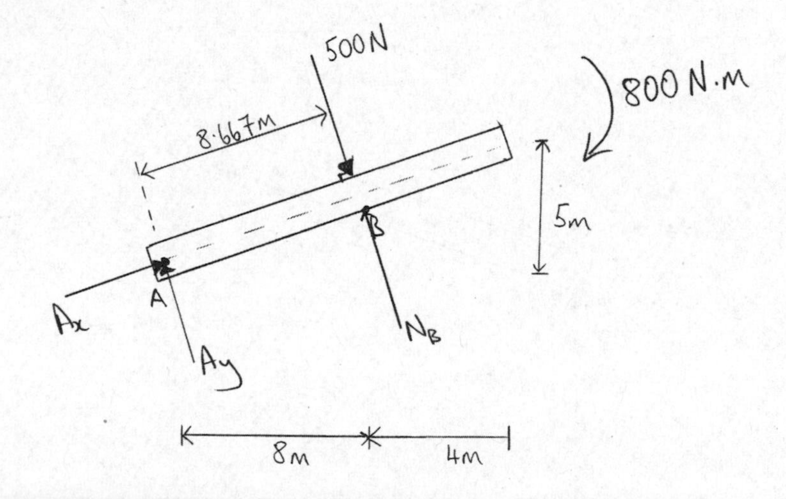

A_x, A_y : effect of pin support on beam
N_B : force of rocker support on beam
800N.m : effect of applied couple moment on beam
500N : effect of applied force on beam

Problem 3.25

Draw the free-body diagram of the vehicle, which has a mass of 5 Mg and center of mass at G. The tires are free to roll, so rolling resistance can be neglected.

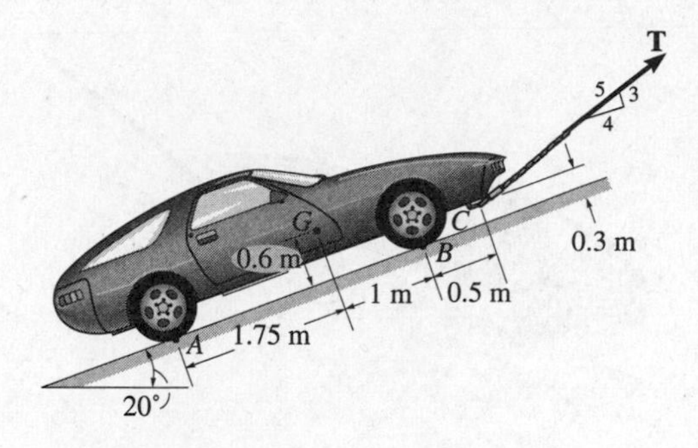

Solution

1. Imagine the vehicle to be separated or detached from the system.
2. There are smooth contacts at A and B. Use Table 2.1 (6) to determine the number and types of reactions *acting on the vehicle* at A and B.
3. The vehicle is subjected to four *external* forces. They are caused by:

   i. **ii.**

   iii. **iv.**
4. Draw the free-body diagram of the (detached) vehicle showing all these forces labeled with their magnitudes and directions. *Assume* the sense of the vectors representing the *reactions acting on the vehicle* (the correct sense will always emerge from the equilibrium equations for the vehicle). Include any other relevant information e.g. lengths, angles etc. which may help when formulating the equilibrium equations (including the moment equation) for the vehicle.

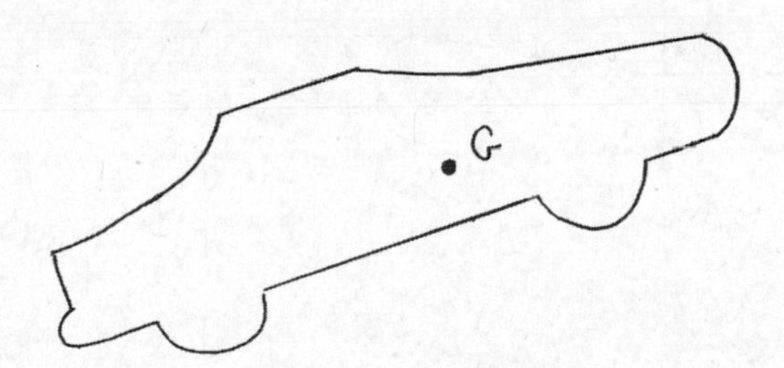

Problem 3.25

Draw the free-body diagram of the vehicle, which has a mass of 5 Mg and center of mass at G. The tires are free to roll, so rolling resistance can be neglected.

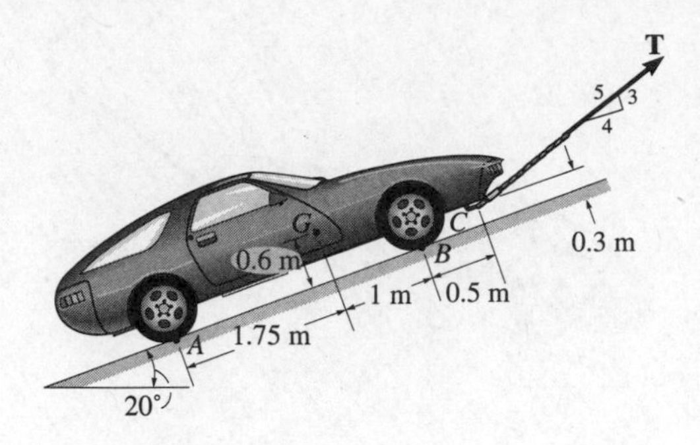

Solution

1. Imagine the vehicle to be separated or detached from the system.
2. There are smooth contacts at A and B. Use Table 2.1 (6) to determine the number and types of reactions *acting on the vehicle* at A and B.
3. The vehicle is subjected to four *external* forces. They are caused by:

   i. The reaction at A

   ii. The reaction at B

   iii. Car's weight

   iv. Force T

4. Draw the free-body diagram of the (detached) vehicle showing all these forces labeled with their magnitudes and directions. *Assume* the sense of the vectors representing the *reactions acting on the vehicle* (the correct sense will always emerge from the equilibrium equations for the vehicle). Include any other relevant information e.g. lengths, angles etc. which may help when formulating the equilibrium equations (including the moment equation) for the vehicle.

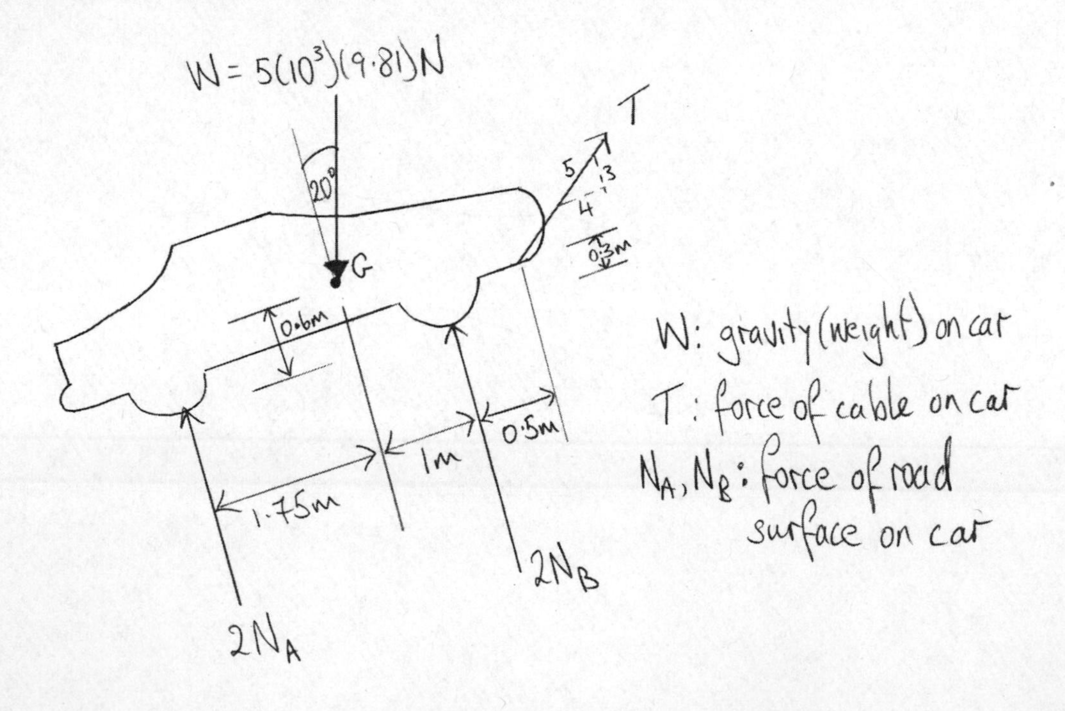

Problem 3.26

Draw a free-body diagram of the crane boom ABC, which has a mass of 45 kg, center of gravity at G, and supports a load of 30 Kg. The boom is pin-connected to the frame at B and connected to a vertical chain CD. The chain supporting the load is attached to the boom at A.

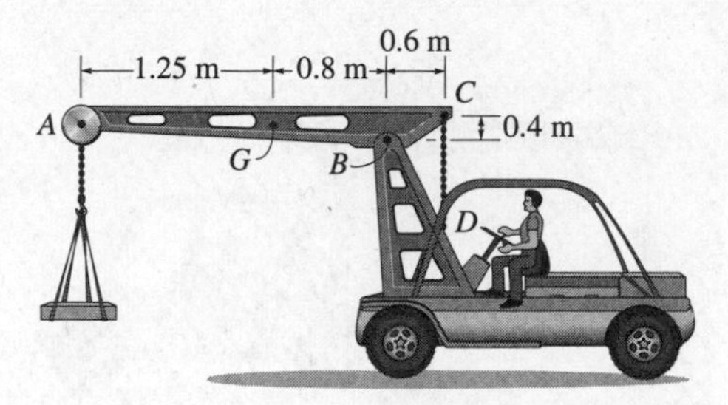

Solution

1. Imagine the boom to be separated or detached from the system.
2. There is a pin connection at B and a vertical chain (cable) support at C. Use Table 2.1 (1) and (8) to determine the number and types of reactions *acting on the boom* at B and C.
3. The boom is subjected to five *external* forces. They are caused by:

   i. **ii.**

   iii. **iv.**

   v.

4. Draw the free-body diagram of the (detached) boom showing all these forces labeled with their magnitudes and directions. *Assume* the sense of the vectors representing the *reactions acting on the boom.* Include any other relevant information e.g. lengths, angles etc. which may help when formulating the equilibrium equations (including the moment equation) for the boom.

Problem 3.26

Draw a free-body diagram of the crane boom ABC, which has a mass of 45 kg, center of gravity at G, and supports a load of 30 Kg. The boom is pin-connected to the frame at B and connected to a vertical chain CD. The chain supporting the load is attached to the boom at A.

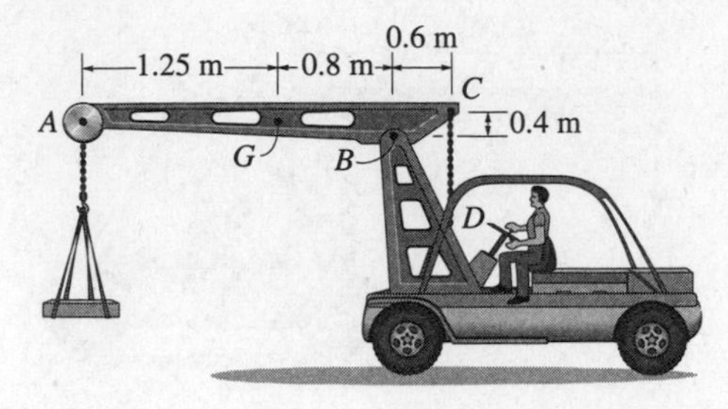

Solution

1. Imagine the boom to be separated or detached from the system.
2. There is a pin connection at B and a vertical chain (cable) support at C. Use Table 2.1 (1) and (8) to determine the number and types of reactions *acting on the boom* at B and C.
3. The boom is subjected to five *external* forces. They are caused by:

   i. & ii. The reactions at B

   iii. The reaction at C

   iv. Weight of boom

   v. Load at A

4. Draw the free-body diagram of the (detached) boom showing all these forces labeled with their magnitudes and directions. *Assume* the sense of the vectors representing the *reactions acting on the boom.* Include any other relevant information e.g. lengths, angles etc. which may help when formulating the equilibrium equations (including the moment equation) for the boom.

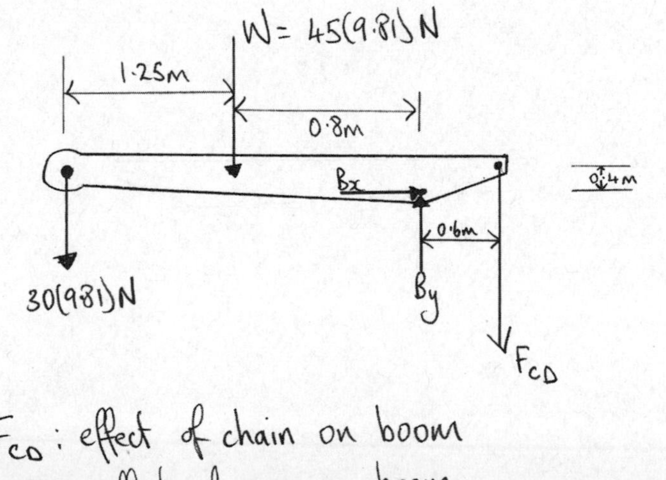

F_{CD} : effect of chain on boom

B_x, B_y : effect of pin on boom

30(9.81)N : weight of load (through chain) on boom

W : effect of gravity (weight) on boom

Problem 3.27

Draw a free-body diagram of the beam. Neglect the thickness and weight of the beam.

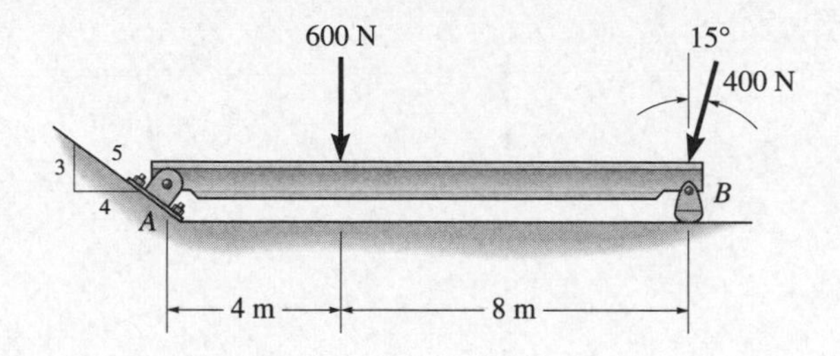

Solution

1. Imagine the beam to be separated or detached from the system.
2. There is a pin connection at A and a rocker support at B. Use Table 2.1 to determine the number and types of reactions *acting on the beam* at A and B.
3. The beam is subjected to five *external* forces. They are caused by:

   i. **ii.**

   iii. **iv.**

   v.
4. Draw the free-body diagram of the (detached) beam showing all these forces labeled with their magnitudes and directions. *Assume* the sense of the vectors representing the *reactions acting on the beam*. Include any other relevant information e.g. lengths, angles etc. which may help when formulating the equilibrium equations (including the moment equation) for the beam.

Problem 3.27

Draw a free-body diagram of the beam. Neglect the thickness and weight of the beam.

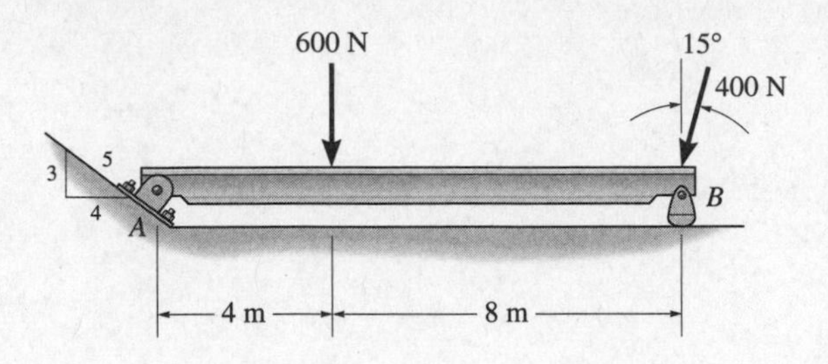

Solution

1. Imagine the beam to be separated or detached from the system.
2. There is a pin connection at A and a rocker support at B. Use Table 2.1 to determine the number and types of reactions *acting on the beam* at A and B.
3. In addition to those shown in the figure, the beam is subjected to three *external* forces. They are caused by:

   i. & ii. The reactions at A **iii. The reaction at** B

4. Draw the free-body diagram of the (detached) beam showing all these forces labeled with their magnitudes and directions. *Assume* the sense of the vectors representing the *reactions acting on the beam.* Include any other relevant information e.g. lengths, angles etc. which may help when formulating the equilibrium equations (including the moment equation) for the beam.

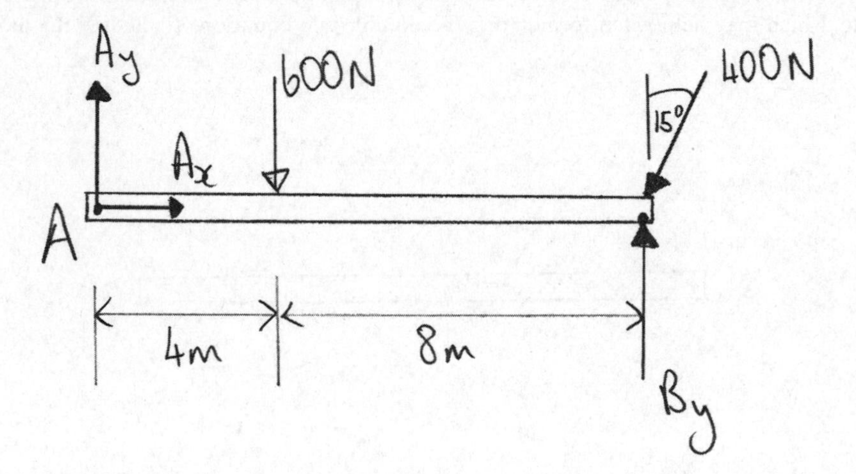

Problem 3.28

The link shown in the figure is pin-connected at A and rests against a smooth support at B. Draw the free-body diagram for link ABC and use it to compute the horizontal and vertical components of reaction at pin A. Neglect the weight of the link.

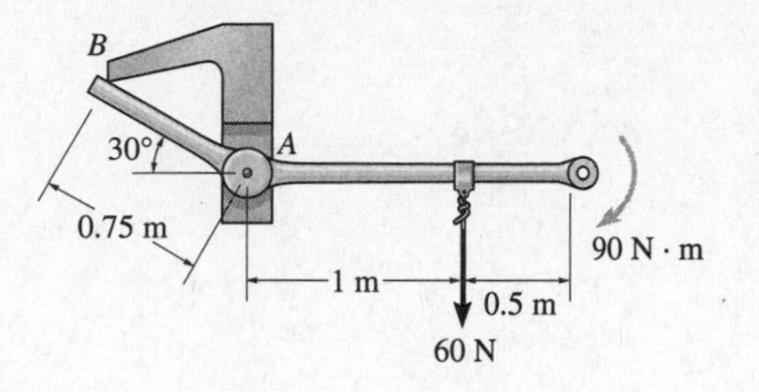

Solution

1. Imagine the link ABC to be separated or detached from the system.
2. There is a pin connection at A and a smooth support at B. Use Table 2.1 to identify the reactions *acting on the link* at A and B.
3. The link is subjected to four *external* forces and one external applied couple moment.
4. Draw the free-body diagram of the (detached) link showing all these forces and couple moments labeled with their magnitudes and directions. *Assume* the sense of the vectors representing the *reactions acting on the link*. Include any other relevant information e.g. lengths, angles etc. which may help when formulating the equilibrium equations (including the moment equation) for the link.

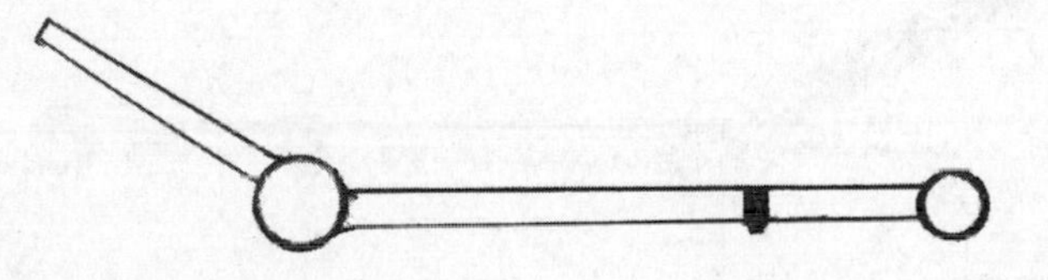

5. Sum moments about A and write down the moment equilibrium equation.

$\curvearrowleft + \sum M_A = 0$:

6. Establish an xy-axes system on the free-body diagram and write down the force equilibrium equations in each of the x and y-directions

$\underset{\rightarrow}{+} \sum F_x = 0$:

$+\uparrow \sum F_y = 0$:

7. Solve the three equations for the required reaction components at pin A:

Problem 3.28

The link shown in the figure is pin-connected at A and rests against a smooth support at B. Draw the free-body diagram for link ABC and use it to compute the horizontal and vertical components of reaction at pin A. Neglect the weight of the link.

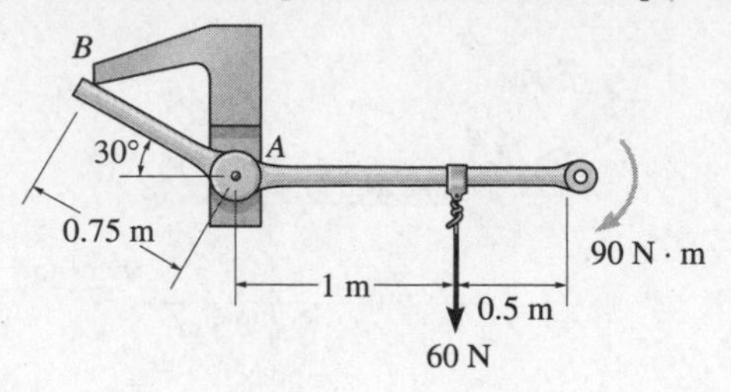

Solution

1. Imagine the link ABC to be separated or detached from the system.
2. There is a pin connection at A and a smooth support at B. Use Table 2.1 to identify the reactions *acting on the link* at A and B.
3. The link is subjected to four *external* forces and one external applied couple moment.
4. Draw the free-body diagram of the (detached) link showing all these forces and couple moments labeled with their magnitudes and directions. *Assume* the sense of the vectors representing the *reactions acting on the link.* Include any other relevant information e.g. lengths, angles etc. which may help when formulating the equilibrium equations (including the moment equation) for the link.

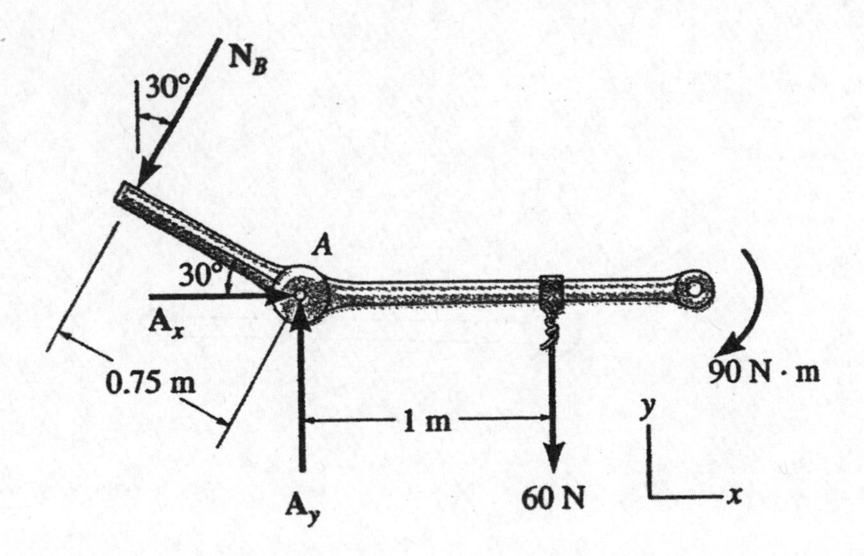

5. Sum moments about A and write down the moment equilibrium equation.

$$\curvearrowleft + \sum M_A = 0: \quad -\ 90N.m\ -\ 60N(1m) + N_B(0.75m)\ = 0$$

6. Establish an xy-axes system on the free-body diagram and write down the force equilibrium equations in each of the x and y-directions

$$\underset{\rightarrow}{+} \sum F_x = 0: \ A_x - N_B \sin 30° N\ = 0$$

$$+\uparrow \sum F_y = 0: \ A_y - N_B \cos 30° N - 60N\ = 0$$

7. Solve the three equations for the required reaction components at pin A:

$$N_B = 200N, \quad A_x = 100N, \quad A_y = 233N$$ **Ans.**

Problem 3.29

A force of 150 lb acts on the end of the beam. Using the free-body diagram for the beam, find the magnitude and direction of the reaction at pin A and the tension in the cable. Neglect the weight of the beam.

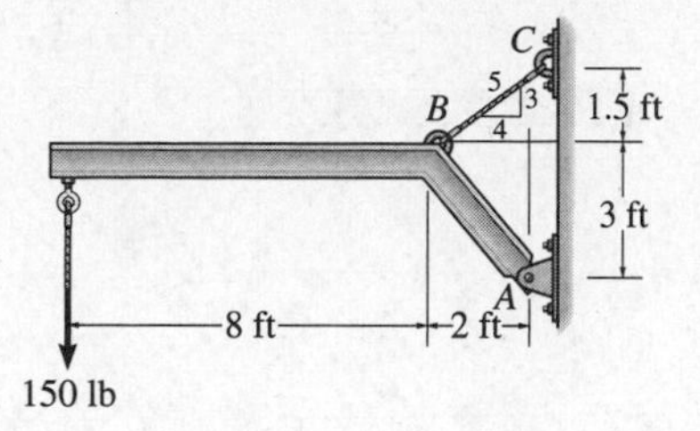

Solution

1. Imagine the beam to be separated or detached from the system.
2. There is a pin connection at A and a cable support at B. Use Table 2.1 to identify the reactions *acting on the beam* at A and B.
3. The beam is subjected to four *external* forces.
4. Draw the free-body diagram of the (detached) beam showing all these forces labeled with their magnitudes and directions. *Assume* the sense of the vectors representing the *reactions acting on the beam.* Include any other relevant information e.g. lengths, angles etc. which may help when formulating the equilibrium equations (including the moment equation) for the beam.

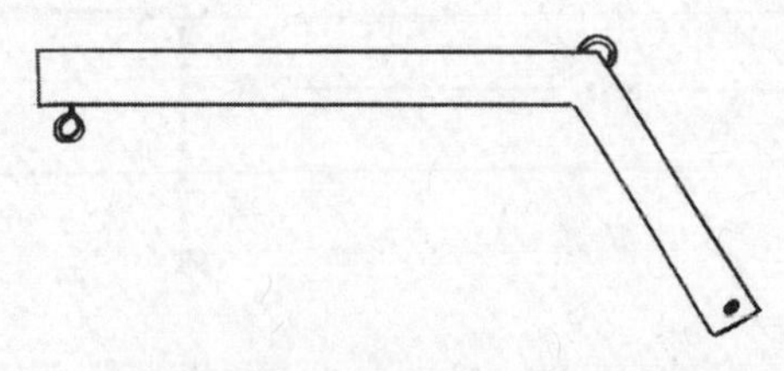

5. Sum moments about A and write down the moment equilibrium equation.

   $\curvearrowleft + \sum M_A = 0:$

   You should obtain the cable tension directly from this equation.
6. Establish an xy-axes system on the free-body diagram and write down the force equilibrium equations in each of the x and y-directions

   $\overset{+}{\rightarrow} \sum F_x = 0:$

   $+\uparrow \sum F_y = 0:$
7. Solve the two equations for the magnitude and direction of the (resultant) force at pin A:

Problem 3.29

A force of 150 lb acts on the end of the beam. Using the free-body diagram for the beam, find the magnitude and direction of the reaction at pin A and the tension in the cable. Neglect the weight of the beam.

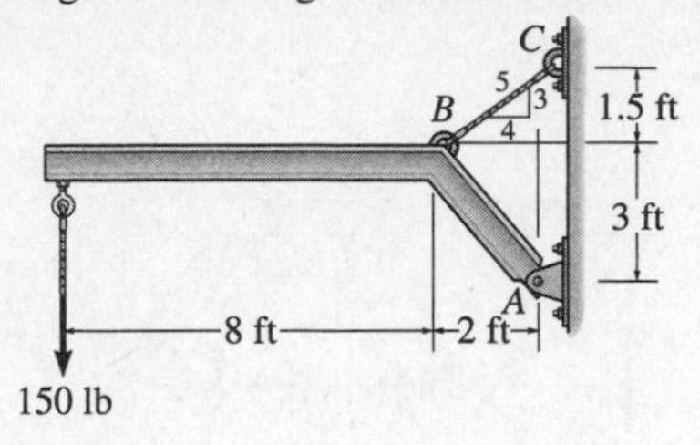

Solution

1. Imagine the beam to be separated or detached from the system.
2. There is a pin connection at A and a cable support at B. Use Table 2.1 to identify the reactions *acting on the beam* at A and B.
3. The beam is subjected to four *external* forces.
4. Draw the free-body diagram of the (detached) beam showing all these forces labeled with their magnitudes and directions. *Assume* the sense of the vectors representing the *reactions acting on the beam.* Include any other relevant information e.g. lengths, angles etc. which may help when formulating the equilibrium equations (including the moment equation) for the beam.

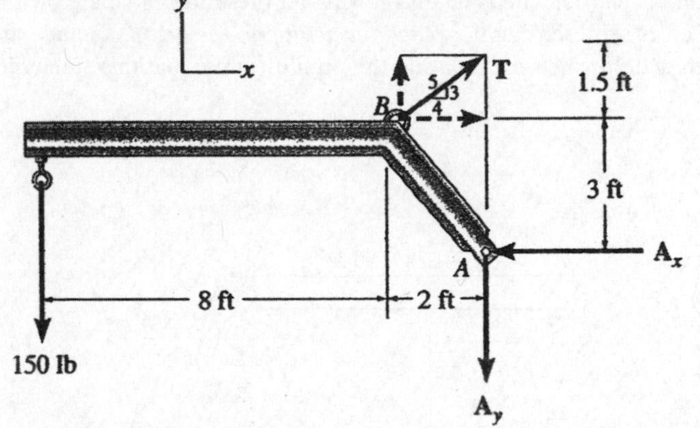

5. Sum moments about A and write down the moment equilibrium equation.

$$\curvearrowleft + \sum M_A = 0: \quad -\left(\frac{3}{5}T\right)(2\text{ ft}) - \left(\frac{4}{5}T\right)(3\text{ ft}) + 150lb(10ft) = 0$$

You should obtain the cable tension directly from this equation:

$$T = 416.7\text{ lb.}$$ **Ans.**

6. Establish an xy-axes system on the free-body diagram and write down the force equilibrium equations in each of the x and y-directions

$$\underset{\rightarrow}{+} \sum F_x = 0: \quad -A_x + \left(\frac{4}{5}\right)(416.7\text{ lb}) = 0$$

$$+\uparrow \sum F_y = 0: \quad \left(\frac{3}{5}\right)416.7\text{ lb} - 150\text{ lb} - A_y = 0$$

7. Solve the two equations for the magnitude and direction of the (resultant) force at pin A:

$$\mathbf{A}_x = 333.3\text{ lb} \longleftarrow, \quad \mathbf{A}_y = 100\text{ lb} \downarrow$$

Thus, magnitude of force at A is $\sqrt{(333.3)^2 + (100)^2} = 348.0$ lb

Direction is $\theta = \tan^{-1}\dfrac{-100}{-333.3} = 196.7^\circ$ (16.7°) **Ans.**

Problem 3.30

The oil-drilling rig shown has a mass of 24 Mg and mass center at G. If the rig is pin-connected at its base, use a free-body diagram of the rig to determine the tension in the hoisting cable and the magnitude of the hoisting force at A when the rig is in the position shown.

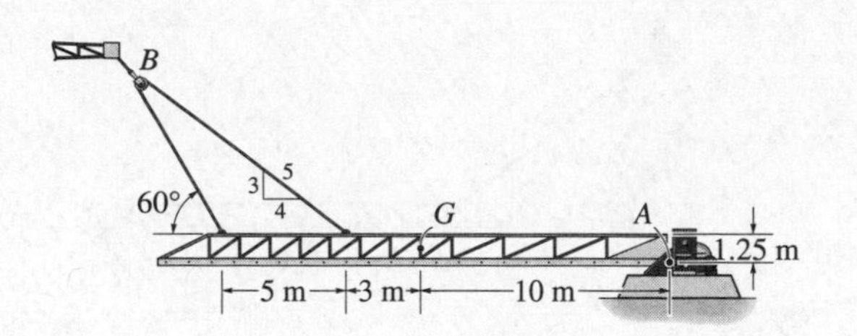

Solution

1. Imagine the rig to be separated or detached from the system.
2. There is a pin connection at A and a cable support at B. Use Table 2.1 to identify the reactions *acting on the rig* at A and B. Note that since the hoisting cable is continuous and passes over the pulley, the cable is subjected to the same tension T throughout its length.
3. The rig is subjected to five *external* forces.
4. Draw the free-body diagram of the (detached) rig showing all these forces labeled with their magnitudes and directions. *Assume* the sense of the vectors representing the *reactions acting on the rig*. Include any other relevant information e.g. lengths, angles etc. which may help when formulating the equilibrium equations (including the moment equation) for the rig.

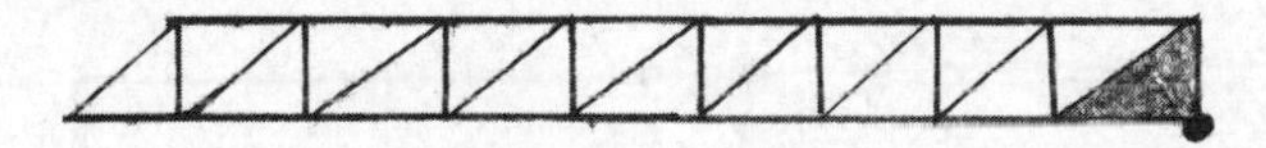

5. Sum moments about A and write down the moment equilibrium equation.

   $\curvearrowleft + \sum M_A = 0$:

   You should obtain the cable tension T directly from this equation:
6. Establish an xy-axes system on the free-body diagram and write down the force equilibrium equations in each of the x and y-directions

   $\underset{\rightarrow}{+} \sum F_x = 0$:

   $+ \uparrow \sum F_y = 0$:
7. Solve the two equations for the magnitude of the (resultant) force at pin A:

Problem 3.30

The oil-drilling rig shown has a mass of 24 Mg and mass center at G. If the rig is pin-connected at its base, use a free-body diagram of the rig to determine the tension in the hoisting cable and the magnitude of the hoisting force at A when the rig is in the position shown.

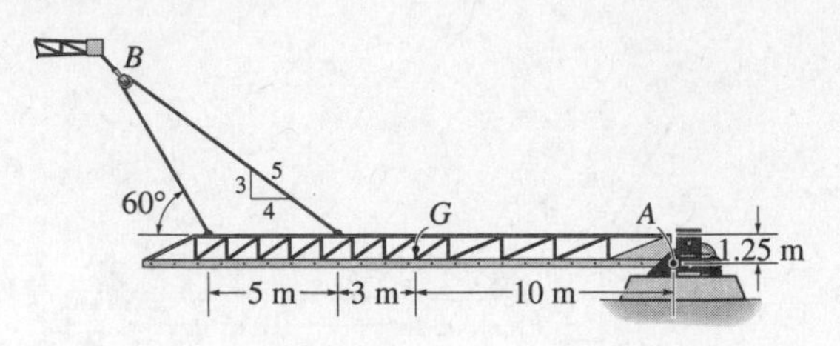

Solution

1. Imagine the rig to be separated or detached from the system.
2. There is a pin connection at A and a cable support at B. Use Table 2.1 to identify the reactions *acting on the rig* at A and B. Note that since the hoisting cable is continuous and passes over the pulley, the cable is subjected to the same tension T throughout its length.
3. The rig is subjected to five *external* forces.
4. Draw the free-body diagram of the (detached) rig showing all these forces labeled with their magnitudes and directions. *Assume* the sense of the vectors representing the *reactions acting on the rig*. Include any other relevant information e.g. lengths, angles etc. which may help when formulating the equilibrium equations (including the moment equation) for the rig.

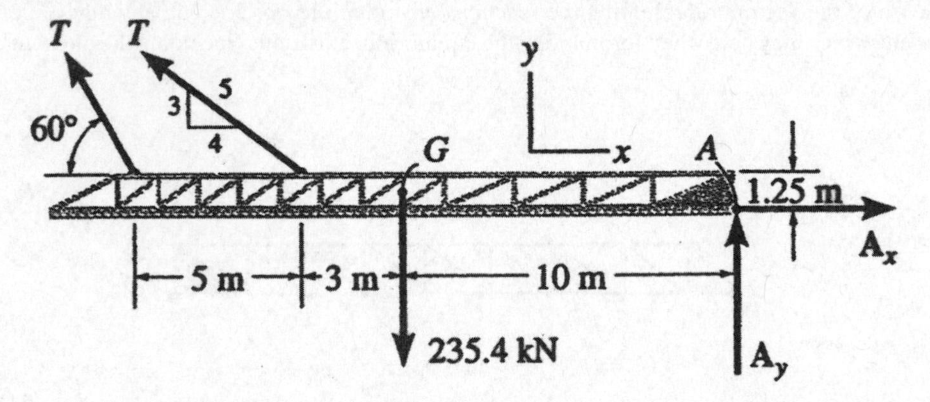

5. Sum moments about A and write down the moment equilibrium equation.

$$\curvearrowleft + \sum M_A = 0:\ (235.4kN)(10m) - \left(\frac{3}{5}\right)T(13m) + \left(\frac{4}{5}T\right)(1.25m) - T\sin 60^\circ(18m) + (T\cos 60^\circ)(1.25m) = 0$$

You should obtain the cable tension T directly from this equation: $T = 108.2$ kN

6. Establish an xy-axes system on the free-body diagram and write down the force equilibrium equations in each of the x and y-directions

$$\overset{+}{\rightarrow} \sum F_x = 0:\ A_x - 108.2\left(\frac{4}{5}\right) \text{ kN} - 108.2\cos 60^\circ \text{ kN} = 0$$

$$+\uparrow \sum F_y = 0:\ A_y - 235.4 \text{ kN} + 108.2\left(\frac{3}{5}\right) \text{ kN} + 108.2\sin 60^\circ \text{ kN} = 0$$

7. Solve the two equations for the magnitude of the (resultant) force at pin A:

$$A_x = 140.6 \text{ kN},\ A_y = 76.8 \text{ kN}$$

Thus, magnitude of force at A is $\sqrt{(140.6)^2 + (76.8)^2} = 160$ kN **Ans.**